SpringerBriefs in Applied Sciences and Technology

PoliMI SpringerBriefs

D1331356

More information about this series at http://www.springer.com/series/11159
http://www.polimi.it

Pelin Arslan

Mobile Technologies
as a Health Care Tool

POLITECNICO
DI MILANO

Pelin Arslan
Mobile Experience Laboratory
Massachusetts Institute of Technology
Cambridge, MA
USA

ISSN 2191-530X   ISSN 2191-5318   (electronic)
SpringerBriefs in Applied Sciences and Technology
ISSN 2282-2577   ISSN 2282-2585   (electronic)
PoliMI SpringerBriefs
ISBN 978-3-319-05917-4   ISBN 978-3-319-05918-1   (eBook)
DOI 10.1007/978-3-319-05918-1

Library of Congress Control Number: 2015957237

Printed on acid-free paper

This Springer imprint is published by SpringerNature
The registered company is Springer International Publishing AG Switzerland

Preface

The aim of this book is to provide guidance in finding the hidden power of interactions and creativity of participation through technologies, in order to construct healthier lives by and for people. This work discusses the importance of healthcare and disease prevention. It offers tools to guide personalized health services to promote healthier lives, and to optimize health management through awareness, motivation, and engagement. The initial part of the book proposes mobile technology solutions, discusses some case studies in health care, and explores various collaborative applications in mobile health to aid in prevention. Furthermore, the conceptual framework was demonstrated with a sample project that had been developed and evaluated through an onsite deployment using participatory tools and strategies proposed in obesity prevention. As a conclusion, the work proposes future directions in participation of everyday health through mobile technologies, starting from personal health prevention and continuing to a systematic approach of a collaborative platform with various stakeholders to cover all influence levels of individual healthier lifestyles.

Healthcare today is in transition toward well-being in the daily lives of individuals. The concept of a healthier lifestyle requires a long-term relation with your everyday health and moves toward a participative approach for chronic disease prevention. Mobile technologies offer new interventions for a more user-centered, socially connected, and economically sustainable healthcare system. With the improvement of social media technologies, they can help to promote social health connections enabling systems that can monitor, track, and respond to changing health status. A major focus of the book aims to provide this conceptual framework with collected case studies and a sample project to understand and contribute participatory processes and tools through effective use of mobile technologies in promoting healthier lifestyles.

The work provides initial background information on mobile health interventions and research strategies to enhance participatory approach in everyday health management and new media applications as social platforms. The book explains the conceptual framework with a sample project, Locast Health Diary, which aims to provide a helpful set of tools for teens at risk for obesity to record their sociopsychological environment and everyday health routines. In more detail, the study discusses the utility of geo-located video diaries and social networks, where social interactions on the Web and user-recorded video diaries create awareness and help subjects to self-reflect on their activities, aiming to think about positive behavior change. The study furthermore evaluates the use of health diary tools for confronting the obesity problem as part of overall prevention for chronic diseases and, as a conclusion, provides a methodological layout for future research directions while discussing participatory tools to be used in other application areas of mobile health.

Conflict of interest. Author reports no conflict of interest.

Acknowledgments

I am devoting my book to my family for their support in all means throughout my educational life.

I would like to express my gratitude and heartily thankful to my supervisor, Fiammetta Costa, for excellent guidance, encouragement, patience, and emotional support that she has given me continuously over the years of my Ph.D. research.

I would like to show my gratitude to Federico Casalegno to provide me the opportunity and his continuous support to realize my studies in his research group at MIT.

I would like to give my special thanks to team members in TeDH, Technology and Design Group at Politecnico di Milano, and MIT Mobile Experience Lab at Massachusetts Institute of Technology for their contribution and their knowledge, critics to my research activities.

This work gives me the opportunity to show my gratitude for the people who stood by my side during these three years for my thesis. I am indebted to many of my colleagues and friends. I thank all the people who provided their assistance to me in form of advice, suggestions, comments, discussions, ideas, any forms of support, and motivation.

Lastly, I offer my regards to all of those who supported me in any respect during the completion of the project.

Contents

Chapter 1
Introduction

1.1 The Context

A combination of Health, Technology and Design is the context of this book. Healthcare is vital for life. It is a complex problem area that needs different approaches to provide solutions from different perspectives. Technology enables us to construct patterns to ease people's lives. Mobile technologies in particular offer opportunities to be socially connected and actively participate to create user content. Design is a process that can make connections. Contemporary design thinking is moving towards a service-oriented and participative approach to providing solutions bridging healthcare, as a problem area, and mobile technology, as an opportunity area. Service design finds new frameworks to intersect with the problem and discover opportunity areas, providing future scenarios for a healthier lifestyle (Fig. 1.1). These scenarios involve participation in everyday health and involvement in a social context for a healthier life.

1.1.1 Healthcare

Healthcare is a complex system in which patients have traditionally not been regarded as a relevant factor within the system. Current solutions are more focused on medical technology and economics aspects to cure diseases and administer their cost values. While new medical technologies will improve the practice of medicine to cure illnesses, it is even more critical to provide technology that promotes a healthier lifestyle and social connections through supportive environments that proactively help people remain healthy, autonomous, and engaged in life. The World Health Organization (WHO) (1948) defines health as "a state of a complete physical, mental and social well-being". By the same token, not only should the

© The Author(s) 2016
P. Arslan, *Mobile Technologies as a Health Care Tool*,
PoliMI SpringerBriefs, DOI 10.1007/978-3-319-05918-1_1

Fig. 1.1 Circles of interest

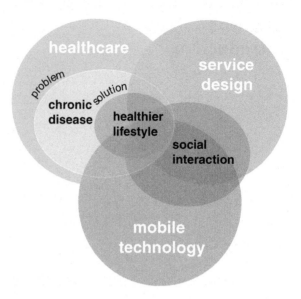

medical situation of the patient be considered but also their everyday environment. This requires an active engagement of patients in their health management since the individual patient is the only person who experiences his or her daily life.

Let us consider an individual's everyday environment and wellbeing. Wellbeing requires long-term management with our own healthcare engaging actively every day. This would bring less cost for healthcare and prevention of chronic disease, which aims to solve prior problems. Technology can help to manage and promote everyday health to digital spaces tailored to physical, cognitive, and social needs of individuals. This enables systems that can monitor, track and respond to changing health status, and create social health connections.

1.1.2 Mobile Technology

Mobile and web applications are becoming increasingly important for our everyday healthcare to support social interactions. Smart phones are now a choice for accessing various applications and services from self-managing our health data to generating collective actions for a healthier lifestyle.

Technologies can be applied to create different forms of social support platforms that allow people to not only store their health data but also generate collective action with other people. This collective action effects individual's health choices in their environment. In their study, Cottam and Leadbetter (2004) show that people who are a part of a community have more access to care and support from friends and neighbors. Social support and a sense of belonging are a vital part of good health. As Bruckner and Bearman (2005) state, sharing information and experiences with other people socially allows them to be motivated emotionally and triggers a modification of patients' behavior. This is an important value especially for chronic disease patients where the treatment is a long-lasting process of the patients' life. Social media promotes these new network structures for mobile communities that are open to new collaborations. Wing and Jeffery (1999) contend in their work that network phenomena might be exploited to spread positive health behaviors, in part because people's perceptions of their own risk of illness may depend on the people around them. It is important to explore new ways of using mobile and social media technologies to promote user-centric solutions to our everyday healthcare and wellbeing.

1.1.3 Design Approach

In order to address social needs of today, design as a discipline has an important role in providing innovative practices in many areas of everyday healthcare and wellbeing context. Within recent economic and social changes of, design is integrating itself into other disciplines to look for new answers and possibilities to satisfy a person's unmet needs.

> On-going shift towards an economy based on services and knowledge, a new vision emerged 'from possession to access' which we may define as the access-based wellbeing, quality of life tends to be related to the quantity and quality of services and experiences (Manzini 2002).

Design thinking moves towards a more service-oriented approach to encourage active participation of users, experts and stakeholders of the system in their problem solving process. Service Design enables and provides methods and tools to create this type of collaboration to confront complex problems on a higher level, since their complexity requires different approaches and competences for a more holistic solution. Sanders and Stappers (2008) explain in their work that the evolution in design research, from a user-centered approach to co-designing, is changing the roles of the designer, the researcher and the person formerly known as the "user". It is a change from a user-centered design process to participatory experiences where the aim is to design a system with multiple different perspectives and a shared goal. Leivrouw (2006) states, "Participatory design is both the means of designing usable and meaningful technologies as well as the outcome of successful systems." Participation and co-creation results in collaborative and distributed solutions tapping into people's perceptions, expectations, desires and motivations. It is also crucial, as Zuboff and Maxmin (2002) argue, that co-creation should provide people with the support they need to follow through on decisions.

1.2   The Problem

Healthcare is one of the most important issues in a human life. There are various problems in today's healthcare such as cost, accessibility to services, lack of healthcare providers and disease-oriented approach of actual systems.

Kung et al. (2008) state that chronic diseases determine seven out of 10 deaths among Americans each year. More than 50 percent of all those deaths each year are due to heart disease, cancer and stroke. For acute illnesses, today's healthcare system answers adequately; however for chronic illnesses, it requires a longer-term treatment with utilization of hospital services, medical sources and high necessity of healthcare providers, which results in high cost healthcare expenditure for both patients and government. Since the complexity of the problem involves more stakeholders within the healthcare system to provide solutions from different per-spectives, it becomes crucial to design participative services.

As described in Fig. 1.2, the alternative solution to a chronic disease problem could be the prevention of chronic disease using mobile technologies as the tool and service design as the strategic approach to sustain a healthy lifestyle. The book further describes this solution set with a community project in the obesity pre-vention area.

1.2.1   Chronic Disease

Chronic diseases are long duration illnesses with generally slowing progression such as heart disease, stroke, cancer, diabetes, obesity etc. Health damaging

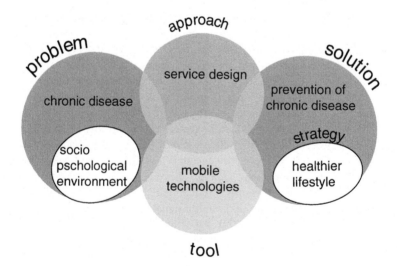

Fig. 1.2 Interrelations between problem area, solution area, tools and strategies

behaviors—particularly tobacco use, lack of physical activity, and poor eating habits—together with environmental problems are major contributors to chronic diseases. However, these diseases are preventable and controllable by daily activities and changing "unhealthy lifestyle choices."

Prevention requires a long-term management of everyday behaviors towards healthier habits such as eating nutritious foods, becoming more physically active and avoiding tobacco consumption. Although the influence of circles in the healthcare system exceeds the individual decisions, it is still the most independent on other factors but the person himself. These self-manageable activities could prevent chronic diseases and therefore also prevent some of the important problems of the healthcare system. Self-managing an activity requires awareness of what you are doing and the mindset of whether what you are doing is right. As prior findings of the focus project of this book, Locast Health Diary, awareness of people's own behavior is the most underutilized health issue. Being conscious of the situation anticipates an active participation in the prevention.

1.2.2 Patients' Socio-Psychological Environment

A sub-problem of coping with a chronic disease is that patients are not regarded as a relevant factor within the cure. Rich et al. (2002) contend that the patient is a network of physical and psychological functions and interacts with physical, biological, social, and symbolic factors. As Helman (1995) noticed, clinicians traditionally focus on the physical and biological aspects. Social and symbolic environments in which patients live and the meanings that patients derive from illness experiences often are not taken into account by contemporary medicine.

There are many influencing factors and clues about patients' lifestyle outside the hospital that are not regarded by the medical experts and that can also contribute to manage patients' life towards healthy behaviors. Medical experts focus on the current physical health data of the patient and mostly do not know or consider their other remaining 23 hours. This socio-psychological information about a patient gives additional information for more complete chronic disease prevention. This type of patients' data is easier to understand and reflect by themselves with respect to biophysical information where patients might need a certain expertise or help by an expert to decode information in their "everyday language." (Everyday language refers to a common understanding of a regular person who does not have any medical background or is not similar with any medical terms.)

Involving the patient in their care and cure process could be a way to obtain socio-psychological data from that patient. This type of data is useful not only for the medical provider but also for the patients themselves. As Goodare and Lockwood (1999) state "Only the patient knows about his or her experience of illness, social circumstances, habits and behavior, attitudes to risk, values and preferences." Obtaining this type of information from patients themselves rather than observing their actions could be another way to active participation of patients in their everyday health routines. Both clinical research and "experiential knowledge" as defined by Caron et al. (2005) are needed to manage illness successfully.

1.2.3 Patients' Involvement in Health Management

Chronic diseases generally cannot be prevented by vaccines or cured by medication, nor do they just disappear. In order to prevent chronic disease problems, a long-term relation with patients is necessary. As Royston et al. (2004) states "The biggest untapped resources in the health system are not doctors but users."

There is often a gap between medical experts and the patient in managing or preventing a chronic illness. An illness-oriented approach allows medical experts to have full control over the patients' health. However, when the patient is not involved in chronic disease prevention, the solutions remain for that specific moment and do not have a long-range impact. We need to maintain a long-term engagement with the patient in order to sustain the solution that we provide to the patient's health. As Wanless (2002) argues, the future of healthcare in our era of chronic disease should turn on the full engagement of people in their own healthcare; the promotion of good health and prevention of illness. In order to create a healthier lifestyle, patients need to manage long-term relations with their everyday activities towards a healthier life. This happens when the patient is informed how to cope with illness in the long run and is taught how to learn to live in a healthy manner.

The use of mobile technologies and social media can help patients manage chronic disease conditions or change behavior in order to follow a healthier lifestyle in many ways such as tracking and storing their health data. It is critical to create platforms to share these data with friends and families, and consult with their healthcare providers.

1.3   The Methodology

This work is based on Frayling's "research through design" model (1993) and explores the practices and processes of design through participation in a project called "Locast Health Diary."

Schön (1983) introduced the idea of design as a reflective practice where designers reflect back on actions taken in order to improve design methodology. Thus the development of design practices is considered not as the objective of the research, but as an integral part of the project. This practice-based approach is a systematic inquiry with systematic reflections that occurs in practice settings. The goal is to move the knowledge derived from creation to research. The main characteristic of this approach is the built-in flexibility of the process.

The research begins with an exploratory process in order to focus on the research problem area through a four-step approach: Recognition, Sourcing, Definition and Selection. This approach helps to narrow down the initial problem area to specific smaller contexts in which "Locast Health Diary" project has been developed (Fig. 1.3). Throughout this exploratory approach, a Research Network (Fig. 1.4) has been created composed of different actors from research institutions, companies and university research centers. The research network has been constructed through expect meetings with clinical psychologists, design thinkers, healthcare providers, community leaders, and program counselors. Different data collection methods such

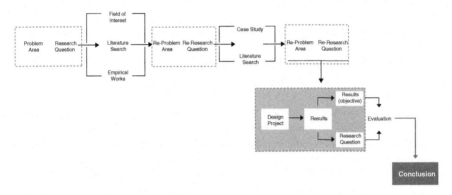

Fig. 1.3 Research process

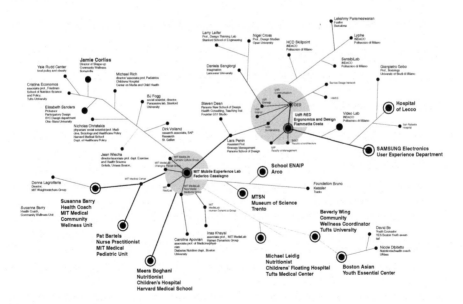

Fig. 1.4 Research network

as discussions, online meetings, and one to one interviews, help to gather information from various perspectives about the problem area and collect case studies in the field. The blue circles are the two research labs, located in two different countries where the expert interviews started to spread out.

The methodology of the research includes several strategies composed of various tools and approaches from different fields that have been used in each stage of research. The four main strategies are: Field Research, Literature Research, Case Studies and Action Research.

Preliminary research is composed of three strategies: Field Research, Literature Research and Case Studies where primary and secondary tools have been used with qualitative data collection tools such as structured and unstructured interviews, questionnaires, participant observation and field research.

Field Research Strategy includes tools such as conferences, research networks, journal surveys on topics regarding "mobile health", "participatory community health", "health services", structured and unstructured interviews of actors included in the research network, which has been used to gather information about the problem area.

Literature Review focuses on three main areas: (a) Mobile Technologies in Healthcare, (b) Design Thinking towards Service Design, (c) Methods, Approaches and Tools for Social Interaction in Health. The review comprises various Ph.D. dissertations and academic articles published in Pubmed, British Medical Journal, Social Science and Medicine, International Journal of Design, Interaction, Co-design and Arts, Clinical Psychology, Social Psychology Journals, Pervasive, Mobile Computing, Persuasive technology papers published in conferences as Connected, AHFE, DRS, DPPI, DESIRE, Cumulus, Ubicomp, Mobile HCI, Mobile Health, Persuasive, HCI, CHI, CSCW, UIST, Oz CHI.

Case Studies Strategy involves case research, case selection, case collection and case evaluation to acquire empirical data and investigate a phenomenon within its real life context. To frame the area of interest and identify the problem, case studies are collected and analyzed from published research projects, seminars and workshops of significant research groups and institutions in industry and university. Case selection is defined according to the different service solution domain in mobile healthcare context and in the research process of each action project.

The main part of the research is conducted through an Action Research Strategy, which includes an action project with three participatory workshops that underpin and guide the research processes. Brydon-Miller et al. (2003) describe how practice and theory can benefit from combining action and research. The "action" took place through the development of design projects. These projects are considered as a terrain or source that inform understanding and guide the evolution of a research process requiring flexibility, intentionality, responsiveness, interventions and participation. The research strategy is planned and conducted in such a way as to disallow the enquiry processes to contaminate the phenomenon under investigation. O'Brien (2001) and McNiff (2002) state that action research is a systematic cyclical method of planning, taking action, observing, evaluating and critical reflecting prior to planning the next cycle of the design process.

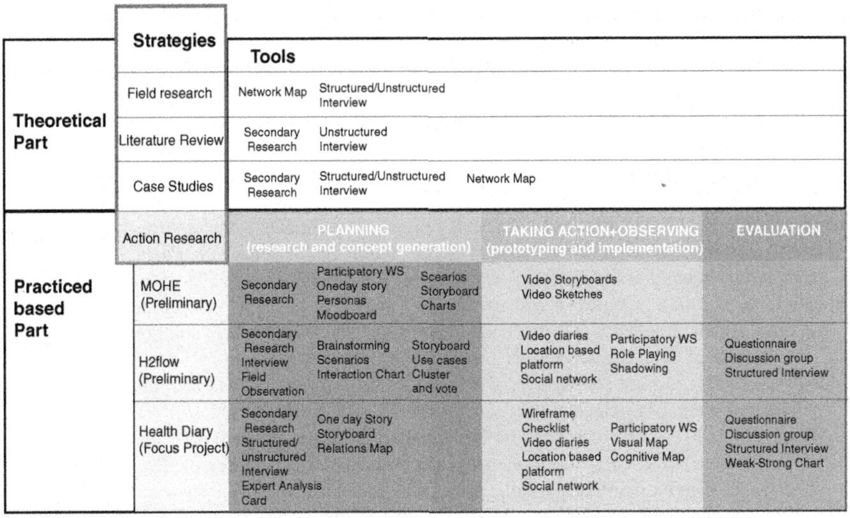

	Strategies	Tools			
Theoretical Part	Field research	Network Map	Structured/Unstructured Interview		
	Literature Review	Secondary Research	Unstructured Interview		
	Case Studies	Secondary Research	Structured/Unstructured Interview	Network Map	
	Action Research	**PLANNING** (research and concept generation)		**TAKING ACTION+OBSERVING** (prototyping and implementation)	**EVALUATION**
Practiced based Part	MOHE (Preliminary)	Secondary Research	Participatory WS, Oneday story, Personas, Moodboard, Scearios, Storyboard Charts	Video Storyboards, Video Sketches	
	H2flow (Preliminary)	Secondary Research, Interview, Field Observation	Brainstorming, Scenarios, Interaction Chart, Storyboard, Use cases, Cluster and vote	Video diaries, Location based platform, Social network, Participatory WS, Role Playing, Shadowing	Questionnaire, Discussion group, Structured Interview
	Health Diary (Focus Project)	Secondary Research, Structured/unstructured Interview, Expert Analysis Card	One day Story, Storyboard, Relations Map	Wireframe Checklist, Video diaries, Location based platform, Social network, Participatory WS, Visual Map, Cognitive Map	Questionnaire, Discussion group, Structured Interview, Weak-Strong Chart

Fig. 1.5 Research methodology diagram

In the action project developed in this research, planning refers to "Research and Concept Development", taking action and observing refer to "Prototyping and Implementation" and evaluating and critical reflecting refers to "Results and Evaluation" (Fig. 1.5).

Prior research conclusions have been collected through interviews and discussions with experts comprising the Research Network, reviewing literature and case studies. For the generation of new ideas and scenarios, various tools such as brainstorming, personas, interaction charts, analysis cards, and mood boards have been used to stimulate collective thinking and concept creation. The implementation phase includes pilot test and deployment through participatory workshops, including tools such as video diaries, social networks, location-based platforms, and cognitive-visual maps. As an Evaluation phase, results have been analyzed with qualitative tools such as questionnaires, focus groups, structured interviews, vote and cluster, and graph analysis.

References

Brydon-Miller, M., Greenwood, D., & Maguire, P., (2003). Why action research? *1*(1): 9–28: 034201[1476-7503(200307)1:1].

Bruckner, H., & Bearman, P. S. (2005). After the promise: The STD consequences of adolescent virginity pledges. *Journal of Adolescent Health, 36*, 271–278.

Caron, J., Lecomte, Y., Strip, E., & Renaud, S., (2005). Predictors of quality of life in schizophrenia. *Community Mental Health Journal, 41*, 399–417.

Cottam, H., & Leadbetter, C. (2004). *Red Paper 01: Health: Co-creating services*. London: Design Council.

Frayling, C., (1993). Research in art and design. Royal College of Art Research Papers 1, vol. 1, pp. 1–5.

Goodare, H., & Lockwood, S. (1999). Involving patients in clinical research: Improves the quality of research. *BMJ, 319*(7212), 724–725.

Helman, C. (1995). *Culture, Health and Illness* (3rd ed.). Oxford: Butterworth Heinemann.

Kung, H. C., Hoyert, D. L., Xu, J. Q., Murphy, S. L. (2008). Deaths: Final data for 2005. *National Vital Statistics Reports 56*(10). Available from: http://www.cdc.gov/nchs/data/nvsr/nvsr56/nvsr56_10.pdf

Leivrouw, L. A. (2006). Oppositional and activist new media: Remediation, reconfiguration, participation. *PDC '06 Proceedings of the ninth conference on Participatory design: Expanding boundaries in design*, Vol. 1, pp. 115–124. ISBN: 1-59593-460-X.

Manzini, E. (2002). Context-based wellbeing and the concept of regenerative solution: A conceptual framework for scenario building and sustainable solutions development. The Journal of Sustainable Product Design *2*(3–4), 141–148.

McNiff, J., (2002). Action research for professional development. Available online at http://www.jeanmcniff.com/ar-booklet.asp

O'Brien, R. (2001). An overview of the methodological approach of action research. In: R. Richardson (Ed.), *Theory and practice of action research*. João Pessoa, Brazil: Universidade Federal da Paraíba (in Portuguese).

Rich, M., Patashnick, J., & Chalfen, R. (2002). Visual illness narratives of asthma: Explanatory models and health-related behavior. *American Journal Health Behavior, 26*(6), 442–453.

Royston, G., Dost, A., Dick, P. (2004). *Healthcare networks and self-care support*. Unpublished paper, Department of Health Operational Research Branch.

Sanders, E., Stappers, P. J. (2008). *CoDesign: International Journal of CoCreation in Design and the Arts, Co-creation and the new landscapes of design. 4*(1), 5–18.

Schön, D. A. (1983). *The reflective practitioner: How professionals think in action*. London: Temple Smith.

Wanless, D. (2002). *Securing our future health: Taking a long term view*. London: HM Treasury.

Wing, R. R., & Jeffery, R. W. (1999). Benefits of recruiting participants with friends and increasing social support for weight loss and maintenance. *Journal of Consulting and Clinical Psychology, 67*, 132–138.

World Health Organization. (1948). Constitution of the World Health Organization. WHO: Basic Documents, Geneva.

Zuboff, S., Maxmin, J. (2002). The support economy: Why corporations are failing individuals and the next episode of capitalism. Allen Lane, The Penguin Press, 2003. ISBN 0713993200.

Chapter 2
Mobile Technologies as a Support Tool for Health

2.1 Mobile Communication and Connectivity for Social Interaction

Mobile technology is very close to everyday lives and mobile devices embody the future of mobile connectivity. Connections between people, information, and place through mobile technologies play an important role designing services that create links and collaborations among different actors of the healthcare system.

Overall, people's behaviors around mobile communication devices have certainly changed over the past decades. Today it is easier to collaborate, share and provide mutual support with peers and networks by using the same mobile devices. Mobile devices are not extra features, but are rooted on some core functionalities and the service experiences that they provide. Even using the same exact hardware of existing mobile phones, it is encouraging to think about new services, redesigning interactions within the same infrastructure.

2.1.1 The Essence of Mobile Technology and Connectivity

Information and Communication Technologies (ICT) have enabled radical changes in our way of finding, storing and using information and systems. Facilities such as Open Source, p2p platforms, Social Networks, Web 2.0 and many others have been dramatically transforming our social infrastructure, economy and industry. Ubiquitous computing, mobile computing, physical computing, pervasive computing, the Internet of things, and haptic computing are post-desktop models of human-computer interaction in which information processing has been thoroughly integrated into everyday objects and activities. All these advancements in technology as Wellman (2001) states change the way people organize and

© The Author(s) 2016
P. Arslan, *Mobile Technologies as a Health Care Tool*,
PoliMI SpringerBriefs, DOI 10.1007/978-3-319-05918-1_2

communicate. Among all these technologies, mobile computing through mobile technologies has a portability aspect.

Portability is one aspect of mobile computing. Since the capabilities of smart phones have evolved, mobile devices are now a common choice for accessing various applications and services. Mobile computing allows us to take a computer and all necessary files and software out into the field. As Feiner (2000), Savidis and Stephanidis (2005) state, the focus in computing and related domains has migrated from the traditional desktop environment into a network of mobile and non-mobile devices. Connectivity through devices everywhere at any time allows for limitless interaction through people, information and places. Connectivity among people fosters relationships and as a consequence creates roots of communities.

2.1.2   Connected Communities

The term 'community' has many meanings. Wilmott (1986) distinguishes three categories of community, which he terms 'place communities', 'communities of interest' and 'communities of attachment'. Each of these typologies is presaged on a commonality. In the first category, it is a shared place of residence; in the second category, it is shared characteristics such as ethnic origin or occupation; and in the third one, it is a shared agreement or compact which brings people together. These three types of community can coincide and in such circumstances, it is suggested, any community feeling would prove to be particularly strong (Crow and Allan 1994: 5). The perfect community is thereby often portrayed as one in which individuals and groups naturally organize themselves to work to actively shape their shared environment, where people act together and '...participate in efforts to address their needs collectively' (Contractor and Bishop 2000: 152). This commonality and participation is what distinguishes a community from many other social groups.

The Internet is a powerful tool with which to create digital communities that enable communication between individuals and groups of people to interact in virtual space. Castells argues (1998) that the Internet is the most appropriate medium of communication in an emerging network society and that it will play an increasingly important role, not only in the way that people choose to communicate with each other but also in the way we form social relationships.

On the Internet, we are often reminded, people can even adopt alternative personas and become whoever they wish to be. We are then not surprised to hear the suggestion:

...through use of, and exposure to, these new technologies, users will adopt new forms of behavior explicitly linked to the technology itself (Rutter 2001: 371).

Wellman is also encouraged by the fact that the constituency from which communities formed via the internet are selected, is in no way bound by geography, but potentially consists of the millions who are online across the globe (Wellman 2000: 15).

Wellman clearly welcomes the growth of such communities and sees in them the potential to develop highly significant relationships. He makes the case that, because these communities are not constrained by geography or chance encounters, they are fully chosen by their members. As Christakis and Fowler (2007) suggested, social distance plays a stronger role than geographic distance in the spread of behaviors or norms associated with such characteristics as obesity.

2.2 Mobile Devices as a Means of Communication Tool

Today, mobile devices are more than just 'personal communication tools'; they are globally networked dominant urban processors. They are not only simple phones with which to call each other but also have their own additional roles, such as tracking health data, positioning location devices, monitoring and measuring conditions, and projecting images. The premise of around-device interaction where it goes beyond the physical constraints of a mobile device, utilizes the space surrounding the device to provide richer input possibilities.

2.2.1 Mobile Devices as High-Capability Processors

Mobile devices, in particular smart phones, are currently undergoing rapid evolution in technology. The improvements in CPU speed, memory capacity, screen resolution and sensory capabilities of these devices have profoundly affected the development of applications. Many mobile devices are equipped with a rich set of sensors, such as GPS, accelerometers, gyroscopes, magnetometers, and distance sensors to create opportunities to implement a much wider variety of applications.

These personalized devices have become a multipurpose tool with built-in cameras, high-resolution color screens, enhanced audio features, sensors and gadgets that aim to provide different functions and various opportunities. However apart from its technological advancement, affordability, accessibility and usability issues are also important factors that enable mobile phones to be used in various contexts. Advances in mobile and sensing technologies provide an opportunity to be used by anyone, particularly non-experts, which mostly include non-technology people, elderly, children and disabled people.

Figure 2.1 shows output devices with on and off board sensors, which are present in recent mobile technologies. On-board sensors are divided into six areas of interest: Information, Light/Vision, Touch, Sound, Orientation/Movement and Location. Off-board sensor components are Bluetooth, ANT+, Near Field Communication and Audio Jack. Output technologies that have been inserted into mobile devices are flashlight, speaker, primary and secondary screen, vibration, and piezoelectric haptic feedback.

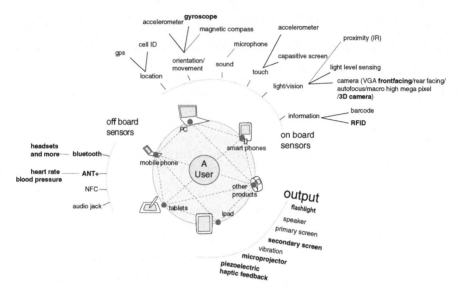

Fig. 2.1 Mobile technologies map

Smart phones combine the functions of a Personal Digital Assistant (PDA) and a mobile phone. They serve as portable phones, media players, cameras, web browsers and GPS navigation systems and have more advanced computing ability and connectivity than a traditional phone. They run different mobile operating systems and also run third-party applications using advanced application programming interfaces (APIs), which can allow them to have better integration with the phone's operating systems and hardware than feature phones.

2.2.2 Mobile Devices as Connected Health Tool

Mobile devices are now able to connect to a variety of sensors and provide personalized information to help people reflect on their activities and improve their health. They can aggregate and visualize everyday health data to encourage interaction, increase awareness and stimulate positive behavior change. For example, monitoring devices or mobile applications such as a weight scale, glucometer, heart-rate sensors, blood pressure monitors, pulse oximeters, implantable cardioverter defibrillators, thermometers, electrocardiographs, peak flow meters, stethoscopes, and pedometers can communicate with another device computer through a Wi-Fi network. They can all send real-time data to a variety of devices. The devices gather information that is then conveyed via communications technologies to a physician, nurse, health coach, emergency medical service or other care provider. The data can provide real-time information necessary for the continuation of care and useful for patient advice, as it can be also stored as electronic health records. Proponents of health monitoring continue to believe that such technology could result in considerable cost savings due to decreased read-missions to hospitals, avoid unnecessary visits to physicians, enhanced medical compliance, and improved communication between patient and clinicians.

Mobile phones have many different usage opportunities. Kindberg et al. (2005) have schematized the motivations driving the use of camera phones for social, personal and affective oriented purposes. People have experienced a considerable diversity of mobile phones; hence it is logical to assume that what people may consider as essential features of a mobile phone should be very diversified as well. Taylor and Harper state (2003) that mobile phones serve as objects play and arti-facts for discussion, and in this role they can function as facilitators of social face-to-face interaction. It has been also shown that mobile phones with multimedia capabilities are used as platforms for expression and creativity. A field study experiment done by Puikkonen et al. (2008) with collaborative mobile video cre-ation demonstrated that mobile phone video can function as a great tool for cre-ativity, as groups of teenagers were inspired to produce and perform in their own mobile mini-movies.

Exchange of multimedia elements creates interactions between the network members. Mobile phones are commonly used to produce these interactions through media content on social networks and other online distribution sites to foster civic media practices. In addition to recording a video through a mobile phone, numerous services allow live broadcasting and archiving of video directly from a wide range of supported mobile devices to the Internet. According to the International Telecommunication Union Report published in 2010, the number of mobile phone subscriptions reached 4.6 billion in 2009, corresponding to 67 per 100 inhabitants globally. In developing countries, the number of mobile phone subscriptions rep-resented 57 %. The high adoption of mobile phones among users has established a foundation for developing potential mobile health solutions.

In Pew's latest study (2010) on 12–17 years old teens, there are interesting insights about teen's use of mobile technologies: According to the research, mobile phones are not just about calling or texting but with expanding functionality, they have become multimedia-recording devices and pocket-sized Internet connected computers. Teens mostly take and share pictures and play music: 83 % use their phones to take pictures, 64 % share pictures with others, 60 % play music on their phones, 46 % play games on their phones, 32 % exchange videos on their phones, 31 % exchange instant messages on their phones, 27 % go online for general purposes on their phones, 23 % access social network sites on their phones, 21 % use email on their phones, 11 % purchase things via their phones.

Organizations have been researching and piloting programs to learn how to impact teen's health behaviors. The Murdoch Children's Research Institute (MCRI) with the Telstra Foundation rolled out a program for teens and young adults that use mobile technology to monitor their mental health including stress levels, coping strategies, dietary and fitness factors. Mount Sinai Hospital has researched and learned that sending text messages to young liver transplant patients can improve their medication adherence and avoid life-threatening complications. Partner's Center for Connected Health recently ran a pilot with pregnant teens with the goal of using mobile technology for outreach. Obstetricians, the doctors who specialize in pregnant women, sent SMS text messages about proper prenatal care and regiments. These messages were educational and prompted many teens to take an action such as attend an office visit. Alex Pelletier, Team Lead, Program and Product Development at Partners states that "We learned that the texting strengthened the relationship. 83 percent of ...reported feeling more supported by the health center case manager".

Puikkonen et al. (2008) content that mobile phones are being used in many ways at the boundary of different levels of privacy. It is often the content that makes the phone feel personal, and handling another's device may be perceived as a violation of privacy (Hakkila and Chatfield 2005). However, Kindberg et al. (2005) and Taylor and Harper (2003) state that showing personal phone content to someone can facilitate ongoing conversation, provoke new topics, and increase group cohesion through enjoyment of shared moments.

Despite the proliferation of mobile and online opportunities, according to the Pew Report (Fox 2009) most people when they have a health question first consult health professionals, friends or family members. There are more than 1,5 million applications available for the iPhone, more than 1,8 million for smart phones running Android operating systems in 2015. There are applications for counting calories and nutrition information; for logging fitness workouts; monitoring vital signs; providing health tips; calculating disease risks or body mass index (BMI); learning about medicines and guiding smoking cessations. According to the Mobile Access Report (Smith 2010), online health information is going mobile and bundled into cell phones, particularly among younger adults aged between 18 and 29 years old. The Pew Internet Project's latest survey (Fox and Duggan 2012) of American's adults, conducted in association with the California Healthcare Foundation, finds that 85 % uses a cell phone. Of those: 17 % of cell owners have used their phone to look up health or medical information, 59 % have software applications on their phones, but only one in four adults actually uses them.

As computers and mobile devices are diffused in our daily lives, it becomes important to understand how people use them to change their behaviors and attitudes. As Fogg (2003) states, a computer can be used as persuasive technology to change attitudes or behaviors of users through persuasion and social influence, but not through coercion. Computer technologies can employ different techniques for changing attitudes and behaviors by increasing self-efficiency, providing tailored information, triggering decision making, and simplifying or guiding people through a process. Computers also work as persuasive media by providing simulated experiences, and persuasive social actors by providing social support, modeling attitudes or behaviors, and leveraging social rules and dynamics.

2.3 Mobile Health

The advancement in mobile technology reflects its solutions in the delivery of healthcare services under the research area of 'Mobile Health'. This area includes various other research themes such as Health 2.0, Medicine 2.0, and e-health through information and communication technologies. Mobile and web services create opportunities to trigger a healthier lifestyle and compromise a more user friendly, socially connected and economically sustainable healthcare system. Mobile applications are being applied in various health areas, such as wellness and prevention, chronic disease management, acute care, post-acute care, rehabilitation and ageing at home. As Christakis and Allison (2006) state, people are connected and so their health is connected. These technological advancements help people to create social interactions to connect people to people and provide learning from others' experiences.

2.3.1 The Definition of Mobile Health (m-Health)

Information and communication technologies have changed the form and quality of the delivery of health-related services, commonly known as e-health. Eysenbach (2001) defined e-health as an emerging field in the intersection of medical informatics, public health and business referring to health services and information delivered or enhanced through Internet technologies.

Mobile Health (m-health) is the subset of e-health referring to the delivery of health-related services via mobile communications technology. Examples of mobile health solutions include patient-provider communication, point-of-care data exchange, remote monitoring of medical devices, public health alerts, patient education and clinical trials information. The first occurrence of the terminology 'm-health' in the literature was in the "Unwired e-med" special issue on wireless telemedicine systems (Laxminarayan and Istepanian 2000). Since then, there has been an increased use of the term, encapsulating various types of healthcare systems. Istepanian and Wang (2003) use the term m-health to relate to applications and systems such as telemedicine, and Istepanian and Lacal (2003) such as telehealth.

Mobile Health could provide ubiquitous, inexpensive and reliable opportunities for the epidemiological transition as also stated by Kahn et al. (2008), which is the increase in chronic disease death and disability together with the limitation of present health systems. In the future, mobile health applications will take advantage of technological advances such as device miniaturizations, device convergence, high-speed mobile networks, and improved medical sensors. This will lead to the increased diffusion of mobile health systems both in clinical and patients' environment and moreover in daily life of individuals who manage their everyday health.

2.3.2 The Emergence of Health 2.0

Mobile applications and web platforms open new possibilities to develop solutions that enable effective healthcare access, delivery and self-management of everyday health.

According to the definition of Hughes et al. (2008), Health 2.0 is the use of a specific set of web tools, used by actors in healthcare including caregivers, patients, and scientists, to personalize healthcare, collaborate with each other and promote health education. A number of concepts related to Health 2.0 include telemedicine, electronic medical records, m-health, connected healthcare, and use of the Internet. These concepts enable new approaches for delivering health services and sharing information in a fundamentally different way.

Web-based technologies and applications in health systems make healthcare information more widely available to different stakeholders in a variety of contexts. Web 2.0 is commonly associated with technologies such as weblogs, social bookmarking, wikis, podcasts, RSS feeds, social software, and web application programming interfaces (APIs). The use of Web 2.0 technology promotes collaboration between patients, their caregivers, medical professionals, and other stakeholders in healthcare and promotes public health. Such tools as blogs, podcasts, tagging, search, wikis use the principles of open source, user content generation and the power of networks.

Web 2.0 has been used to stay informed in a particular field through RSS, podcasts and research tools (Giustini 2006); provide medical education for doctors, public health professionals and the general public (Crespo 2007); enable collaboration and practice where Web 2.0 tools are used in daily practices by medical professionals to find information and make decisions (Tan and Ng 2006); share data for research for disease specific communities with rare conditions and aggregate data on treatments, symptoms, and outcomes to improve their decision-making ability and carry out scientific research such as observational trials (Fros et al. 2008).

Hughes et al. (2008) have found six major themes reviewing recent literature in Health 2.0 that include (i) the participants involvement; (ii) the impact on different collaborations and practice; (iii) the ability to provide personalized healthcare; (iv) the use in medical education; (v) associated methods and tools; (vi) privacy and ownership issues with Medicine 2.0 generated information.

In the past, only professional caregivers were able to access medical information and patient records. De Haes argued (2006) that patients were not given some important facts or more helpful knowledge. However, Health 2.0 provides more direct access to health information and helps patients to understand a great deal about their diseases. Digitalization helps healthcare provider connect not only to everything they need to know about patients but also can share information with other healthcare providers who have treated similar disorders and may have the most recent knowledge. Dedicated social networking such as Sermo, SocialMD, and Ozmosis are alternative ways of sharing and collecting data respecting to traditional clinical trial approaches with cognitive questionnaire, individual expert analysis, and quantitative data.

Professionals have now advanced to using media tools such as blogging, where they share their experiences in the form of case studies, share their insights into various diseases, discuss common healthcare issues, and offer simple remedies for professionals with various degrees of expertise.

However there are constraints using these platforms. For example, Hughes et al. (2008) have explored a number of privacy and ownership issues. During research network meetings, the associate director of a mobile strategic group of a pharmaceutical company stated that, in pharma companies, video intervention can be useful in facial expressions analysis. It can, for instance, quantify the strength of pain through a ten-scale face expression, analyzing the patient's emotion through face recognition technology. However one of the constraints of video intervention is a privacy issue for the patient, as well as bureaucratic barriers securing confirmation from other organizations. Moreover, healthcare providers must concern a number of privacy issues when involving their patient in any kind of research experiment as the subject of a study.

In response to the desires of many healthcare professionals, patients and other involved individuals, Health 2.0 is evolving quickly to embrace new technology and new services. However, already there are signs of Health 3.0, which leverages the use of elements of the Semantic Web such as location awareness, the emerging Internet of Things and embedded sensors. In a Health 2.0 and 3.0 setting, there is an increased participation of the patient rather than a traditional model of medicine where physicians have full control of the patients' health data. Such models work better in acute cases when the medical condition is not that serious. However, in the case of complex chronic diseases, psychiatric disorders, or diseases of unknown etiology, patients were at risk of being left without well-coordinated care because data about them was stored in a variety of disparate places and in some cases might contain the opinions of healthcare professionals that were not to be shared with the patient.

The Health 2.0 setting provides patients with the following services: A patient goes to see their primary care physician with a presenting complaint, having first ensured his own medical record was up to date via the internet. The treating physician might make a diagnosis or send for tests, the results of which could be transmitted directly to the patient's electronic medical record. If a second appointment is needed the patient will have had time to research what the results might mean for them, what diagnoses may be likely, and may have communicated with other patients who have had a similar set of results in the past. On a second visit, a referral might be made to a specialist. The patient might have the opportunity to search for the views of other patients on the best specialist to go to, and in combination with their primary care physician decides whom to see. The specialist gives a diagnosis along with a prognosis and potential options for treatment. The patient has the opportunity to research these treatment options and take a more proactive role in coming to a joint decision with their healthcare provider. They can also choose to submit more data about themselves, such as through a personalized genomics service to identify any risk factors that might improve or worsen their prognosis. As treatment commences, the patient can track their health outcomes through a data-sharing patient community to determine whether the treatment is having an effect for them, and can stay up to date on research opportunities and clinical trials for their condition. They also have the social support of communicating with other patients diagnosed with the same condition throughout the world. The products associated with mobile phones provide emotional and clinical support through wireless multi-sensorial sensor technology.

Patient-driven healthcare services are emerging to supplement and extend traditional healthcare delivery models. These services can be characterized as having an increased level of information flow, transparency, and collaboration. These types of services enhance customization, patient choice and responsibility taking as well as quantitative, predictive and preventive aspects. Health 2.0 will be accentuated by disruptive decentralization of medical devices and diagnostic equipment. The cost and inconvenience of these centralized solutions create impetus for innovators to make new products and services, which are simpler and affordable. Designing healthcare products and services in Health 2.0 can produce a vehicle to translate stakeholders' explicit and latent needs into functions, processes and forms and construct an optimized balance among various users.

2.3.3   Future Product Service System in Mobile Health

The rapid advances in information communication technology (Godoe 2000), nanotechnology, bio monitoring (Budinger 2003) mobile networks (Olla 2005), pervasive computing (Akyildiz and Rudin 2001), wearable systems, and drug delivery approaches (Grayson et al. 2004) are transforming the healthcare sector into a decentralized system.

While new medical technologies are improving the practice of medicine, it is even more critical to create supportive environments, technology and platforms that proactively help people remain healthy, autonomous, and engaged in life e.g., such as places tailored to unique physical, cognitive, and social needs of individuals. Appropriate design could provide guidance to enhance user's self-management, consolidate therapeutic relationships, and improve healthcare outcomes. Systems through mobile technologies can monitor and respond to changing health status to promote health and social connections. Distributed resources need to be co-created to address particular needs and circumstances of individuals and communities. There are two indisputable facts about our future mobile devices: They will be equipped with more sensing and processing capabilities and they will also be driven by an architecture of participation and democracy that encourages users to add value to their tools and applications as they use them. Mobile phones are tracking devices that reveal much about our lives. The notion of what and to whom the data was provided threatens the privacy of individuals and causes trust issues regarding provided services.

Mobile health applications play a large and important role in shaping the future of the healthcare system. Some examples in this domain are self-tracking applications to track health parameters such as weight, food, exercise, blood pressure on a daily base; disease management software to record and monitor care; emergency care applications for the elderly, medical alarm systems to reduce the risk of living alone (Lifeline); early detection of development problems of a newborn; motivation of people to have a more active lifestyle (Directlife). These products and applications motivate behavior change by encouraging physical activity, changing diet and quitting smoking. Patients can keep track of their blood pressure, weight history and monitor the progress of their rehabilitation program both on their mobile phone and web-based interface on their computer.

Mobile technologies are accelerating a larger trend of care moving out of hospitals and traditional clinical settings to new venues of health scenarios. As Stachura and Khasanshina (2007) state, Telehealth is a service for wellness and prevention, helping to engage patients in their own health management, and ensuring the continuity of care and improved outcomes. In order to manage chronic diseases successfully, patients need to collect their health data and consult with healthcare professionals to modify their behavior. Eysenbach (2001) has shown that to sustain healthy behaviors, consumers usually need a 'nudge' and motivational support. Small devices for monitoring activities and uploading data to a website where users can track their exercise, calorie consumption and other metrics can help this purpose.

Mobile devices are now able to connect to a variety of sensors and provide personalized information to help people reflect on and improve their health. For example, pedometers, heart-rate sensors, glucometers, and other sensors can all provide real-time data to a variety of devices; mobile devices can aggregate and visualize these types of data to encourage interaction, increased awareness and positive behavior change. While patients and health professionals are enthusiastically using Telehealth solutions and millions of people have downloaded smartphone applications to keep track of their health and wellbeing, digital healthcare has yet to reap its great potential to improve healthcare and generate efficiency savings. Mobile applications connect patients with chronic conditions to caregivers, health coaches, friends and families, communities and interest groups on a continuous basis. But also for people who are not considered as patients, use of many applications focused on wellness, fitness, and nutrition is increasing. Active health conscious individuals are eager to use m-health technologies in everyday activities.

Mobile technologies enable many use cases such as providing real time modification to medication regimens, adjusting lifestyle behaviors, and reminding patients to stick to therapeutic plans in chronic disease management; acute care, post-acute care, rehabilitation, prevention of unnecessary hospital readmissions, aging at home cases as personal emergency response, sensors and broadband at home for healthy and safe aging. Sensor and wearable technologies combined with mobile communication can be useful to track and share various health measurements: a textile product temperature changing sleep suit based on the signals of a baby's fever; Wi-Fi enabled sensors can transmit exercise and psychological signals to rehabilitation specialists or brief healthcare questions can be answered via real-time social media channels.

Self-tracking applications provide information that professionals need in order to develop effective strategies to achieve goals. In some cases the only purpose of a self-tracking tool is for monitoring real-time calorie balance (Tsai et al. 2007). In other cases it might also cultivate behavioral change using emotional or social mechanisms such as in Fish'n'Steps application which attempts to promote a higher level of physical activities by mapping one's steps measured with a pedometer to the development of an animated character, perhaps a fish in a fish tank (Lin et al. 2006). In this kind of application, the user benefits both individually and collectively. Individually, these self-tracking applications create self-awareness and the ability to make better decisions to more easily achieve their goals; whereas collectively, these applications allow social interaction and influence for the purpose of transformative change.

In today's society, the commodity and spread of technology has changed dramatically the way people interact with healthcare. 'Online Health Search 2006', a Pew research report, showed that the Internet is often utilized as a first resource for medical information. Most people start their online health research with a Google search before considering constitutional entities such as the Mayo Clinic or the National Institute of Health (NIH), as stated by the study of Hawkes (2005). Although it is essential to verify health information source and quality, today hundreds of websites and mobile applications are providing medical references for topics like anatomy, first aid, drug-related conditions, but also ideas, desires and experiences of people.

Mobile phones are providing people with a core piece of hardware to build upon for functionality. In addition to hardware, there are external hardware and other device capabilities that extend the usage area of mobile phones in everyday health scenarios. For example, an external hardware used to create iStetoscope, a mobile stethoscope that allows listening to the internal sounds of a human body, was developed by Peter Bentley at University College of London together with Camera Culture Group from Massachusetts Institute of Technology (MIT), iStetoscope uses the combined output of iPhone's microphone, motion sensors and camera to get a much more accurate reading than a simple acoustic stethoscope, then it shows the heart's form on the iPhone's display. Another example is a mobile microscope developed by Aydogan's group at NanoSystem Institute in University of California, Los Angeles (UCLA), where the device uses mobile technology to test for infectious diseases in developing countries.

All these mobile devices and accessories that add additional features lead to the decentralization of healthcare. Health platforms provide status graphs, comparisons with goals and norms, and alerts when things change suddenly or move toward unsafe levels. A research project developed by SAP Research Group (2011) relates to building a patient community consisting of an application where patients can scan doctors and drugs through near-field communication technology (NFC), and are subscribed to a specific community where patients that have the same indication (e.g. blood pressure) can communicate with each other and doctors can communicate with their anonymous patient to provide information. An iPhone application developed by Liu et al. (2011) helps adolescent migraine patients and their counselors to manage migraine. This application provides on-demand access to audio, video and animated instructions to guide the learning and application of behavioral migraine management skills as well as a headache diary that allows adolescents and their counselors to monitor key migraine and medication use variables.

The phone is not just an on-hand computer; it also collects information automatically through its sensors such as GPS, microphone, accelerometer, clock or wearable sensors that can send their data directly to the phone. This creates a body area network. Once reaching the mobile phone, captured physiological metrics can be sent off into the cloud and then to the physician. It can also be analyzed and sent back to a patient. With increasing computational intelligence and real-time information flow, applications will soon influence decision-making.

There are many examples of health-related social media platforms in web and mobile applications that decentralize healthcare services. HealthBuddy, Fitbit, Sleep Cycle and Lark use sensors to measure exercise, weight, and sleep patterns; mobile-based services like Mobile care Monitor, and Welldoc also measure such parameters. Web applications such as Daytum and Open.Sen.se allow people to integrate data from a variety of sources to use visualizations and mash-up tools to find meaning in the data. Weight loss communities, activity-based social networks such as Strava, helps patients to collect and upload vital signs such as blood pressure, weight, and qualitative assessments about pain and mood.

These platforms provide feedback in response to the data received and collected, remind patients about their care and upload vitals to a central platform where software can help medical experts to decide their medical care or determine if it may require in-person intervention.

The most commonly noted weakness of mobile existing applications is the lack of focus on specific target end-users. This could result in the inability to thoroughly identify what users want from a design, how and when they will use it, what makes them want to use it, and the most important of all how to personalize the applications for each individual. There is strong evidence that behavioral change needs to be tailored to match an individual's needs and characteristics in order to be effective and sustainable. Lenert and Kaplan (2000) suggested a more comprehensive approach towards all dimensions of an individual's life to engage users in the long term after an initial period of interest. A comprehensive approach is only feasible when information regarding users is obtained and understood thoroughly. Furthermore, by providing specific target end-users from the beginning, relevant stakeholders could be identified and potentially involved.

Another weakness of existing mobile phone applications is their failure to incorporate relevant behavioral changes theories into the design and development of mobile phone applications. Although incorporation of relevant theories into applications does not necessarily guarantee the success of applications in inducing changes of behavior, there is a strong likelihood that such a system would likely be more capable in achieving this through design. There is also a commonly noted weakness with regard to how the existing applications evaluated their effectiveness in inducing change of behavior. The main issue is the duration of the evaluation period, ranging from a few days to three months. This is considerably shorter than Prochaska and Velicer's (1997) recommendation of a minimum of six months to observe permanent behavior change. The short period of evaluation eventually results in lack of understanding on why or how the mobile phone application could induce behavioral change.

2.3.4 Stakeholders of Mobile Health

Healthcare is a complex system with varieties of stakeholders centered on a user who wants to be cured as a patient or wants to have a healthy life as an individual. Hill (2007) states that innovation is occurring in more venues, not only governmental and industrial research labs but increasingly at technology companies, startups, and small-team academic labs and by creative entrepreneurs and individuals.

In a complete healthcare setting, some of the stakeholders, as seen in (Fig. 2.2), are: Users, online-offline communities, healthcare delivery organizations, healthcare professionals, health insurance companies, manufacturers, service-content providers, mobile carriers, societies and government.

End users of mobile health include not only patients but also healthy people following WHO's definition on health. Online and offline communities are groups of interacting individuals for a common goal. (i.e., PatientsLikeme, Patients for Patient Safety) Healthcare Delivery Organizations deliver professional healthcare in a systematic way. Healthcare professionals as medical experts provide healthcare services and develop collaborative partnerships with patients. Health plans provided by insurance companies work collaboratively with healthcare delivery organizations and clinicians for value-based reimbursement. Manufacturers are producers of medical and health-related products for individuals, and healthcare providers aid them in their activities for health. Service and content providers are specialized for health and healthcare (i.e., Microsoft HealthValut, Virgin HealthMiles). Such carriers are providers of networks for wireless data transport between wireless devices. The societies are communities, organizations who actively participate in efforts to improve healthcare (i.e., WHO (World Health Organization), NHS (National Health Service)). Governments are organizations that act as the governing authority of a political unit. As seen in the stakeholder map, each stakeholder has influence on the user, among each other, and the healthcare system as a whole.

According to the Institute for The Future (2010), this transforming healthcare ecosystem extends beyond healthcare providers and patients to related businesses around them. Besides medical products, a number of applications using mobile technologies have been introduced in the healthcare market. Not only manufacturers but also software developers and mobile carriers have been transformed as stakeholders of the healthcare system. Mobile health creates distributed and

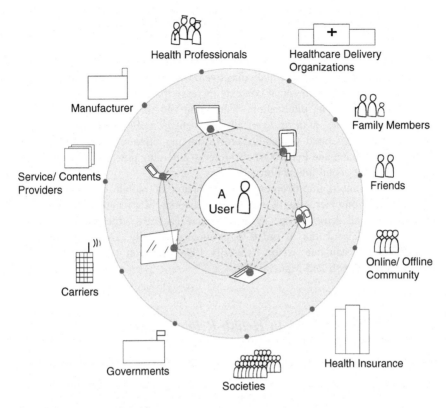

Health Professionals

Healthcare Delivery
Organizations

Manufacturer

Family Members

Service/ Contents
Providers

A
User

Friends

Carriers

Online/ Offline
Community

Governments

Societies

Health Insurance

Fig. 2.2 Healthcare stakeholder map

integrated healthcare and a continuous care model driven by advanced smart systems that help people sense and understand our actions and environments. As IBM suggests in Healthcare 2015: Win-win or lose-lose? (2006), patients are recommended to learn about health, responsibility for living a healthier lifestyle, and demand new delivery models and coordination of care. The front line of healthcare is not where professionals share their knowledge with patients but where people look after themselves. People will only feel empowered to participate in the creation or uptake of a new service if it speaks to them in a language they understand and in a style that is found to be friendly, appealing and is useful for them.

The evolution of social media contributes to this dynamism of participation and coordination of healthcare stakeholders. Moreover, social networks are proliferating and expanding the exchange of information and impacts on healthcare. Sarasohn-Kahn (2008) argues that there is a need for services that "knit" communities together to enable health consumers to move seamlessly and efficiently through networks without having to be a member of all the groups that pertain to their illness or interest. Collaborations between social media and health stakeholders could benefit patients and, at the same time, disrupt traditional relationships between providers, suppliers and consumers.

2.4 Opportunities of Participation in Mobile Health

Participation is both a means of designing usable meaningful technologies, and the outcome of successful systems and collective co-creating tools. Tools and practices of mobile technologies in social contexts increase the capacity for personal communication, production, publication, distribution and sharing. New forms of participation are enabled as a result of advancement in mobile technologies. Terms such as "user generated content", "crowdsourcing" (Howe 2008) and "citizen media" (Trogemann and Pelt 2006) also refer to emerging practices through mobile technologies for social interaction.

Due to advanced improvement and dispersion of mobile and Internet technologies, everyone increasingly uses social networks. Many authors note that it is important for all actors in the system to have an active role participating and enabling social networks in order to promote a socially, environmentally and economically sustainable system (Castells 1996; Tapscott and Williams 2007; Benkler 2006; Jégou and Manzini 2008).

2.4.1 Social Networks in Health-Related Context

Social networks have become a powerful tool for bringing people with similar interests together to interact and share information. Millions of people every day send, upload and download multimedia elements such as text, photo, video and share it among community members or shared resources through these platforms. In a healthcare setting, social networks play a crucial role in providing interaction and exchange of information between people, healthcare providers and other stakeholders of our healthcare system. A health condition is a particularly strong affinity and the collective learning and experience of others can be leveraged and shared to help individuals make decisions. Health social networks such as PatientsLikeMe, CureTogether, DailyStrength, MedHelp, HealthChapter, MDJunction, Experience Project, Peoplejam, and OrganizedWisdom are primarily directed at patients but also caretakers, researchers and other interested actors.

A health social network is a web platform that allows people to find health resources at a number of different levels. According to Swan (2009) these levels range from a basic tier of emotional support and information sharing, question and answer with physicians, quantified self-tracking to clinical trials access.

These type of social networks not only provide patients the potential to find other people in similar health situations but also allow people to share information about their conditions, symptoms and treatments. The largest and best-known health social network is PatientsLikeMe, which started in 2004 based in Boston, USA and had, as of April 2015, more than 325,000 patients.

To date, health social networks have been focused mainly on medical conditions for which cures are sought, although some websites have user communities for healthy living. Some health social networks allow patients explain their thoughts, challenges and rewards as they cope with various health issues (i.e. aids, alzheimer, autism, eating disorder, epilepsy, migraine, cancer, parkinsons). Some platforms enable audio and photo sharing, and others such as sports tracker, Hello health, Collabrhythm (patients become more active participants in their care), and Wellness (open mobile network for activity), Lifelogging, (Sensecom) Calming Technology Group at Stanford University (Moraveji et al. 2011) focus their research on technology that enables calm states of cognition, emotion and physiology for better human health and productivity. Stress management projects such as using social networks to remove the stress of eating breakfast every day, a method of pacing people's breathing while they work on their computer, an SMS-based method of reducing student stress of stepping out of a lecture, continuous augmentation of self-regulation are other types of reasons for individuals to use these health social networks.

Social networks in healthcare is an important phenomenon that not only provides platforms to share data but helps users to connect with their parents, family and friends to get an emotional support during and after treatment. Social networks influence health behaviors. As Smith and Christakis (2008) state, "Social support and a sense of belonging, being part of a social network, are a vital part of good health. People who are part of social networks are generally healthier than people who feel isolated. People who are part of a community have access to care and support from friends and neighbours. They also tend to have higher self-esteem and confidence from a sense of social belonging. Collaboration is vital for people to share and spread ideas and know-how". Centola (2010) in his experimental study shows that people with dense social networks are more likely to acquire new health practices.

Online social networking services are increasingly accessed through mobile devices equipped with location sensing technology. As Langheinrich (2009) states, sharing real-time location information raises important privacy concerns. As discussed in the Wagner et al. (2010) study, there is an increasing amount of work understanding user's location-privacy needs in ubiquitous systems including diary studies (Barkhuus 2004).

Low-cost technology developments in new media platforms allow creation of user content and annotation of this content such as geo-location. As Hemment (2006) states, mobile phones, which are now frequently used as content producers and consumers, are utilizing geo-location for mobile games, travel guides, and participatory art installations. This allows location-aware participatory media platforms for making the user the most highly involved actors within the participatory culture. Peltonen et al. (2007) state platforms that encourage active participation in media production create user experience in which a participant becomes increasingly aware and focused within the situation. As Rheingold (2007) states, these characteristics give rise to the opportunity to use locative media as a supplementary tool for education around complex social topics in areas such as sustainability, community, and civic engagement.

All those applications and platforms represent examples of "augmented local context", where they provide individual information about people and support social interaction at the local level. However, such applications are usually targeted to a younger user group who spend much of their time in front of a computer (whereas elderly people are usually unfamiliar with computers) or people that use more sophisticated communication tools, such as Internet and GPS-enabled touch screen phones.

2.4.2 Participatory Culture Through Social Media

Advancement in mobile technologies and social media allows the creation of new tools for participatory experiences. Information and communication technologies change the way in which we live learn and produce. Designers need to gain the knowledge, processes, or tools to deal with the unfolding of the interactive flow of information.

Health behavior is a social behavior. Some health social networks emphasize social connection and support (i.e., Daily Strength) and some others provide general health information (i.e., Organized wisdom) that patients may be able to enter qualitative and quantitative data about their own conditions, symptoms, treatments and overall experiences. Emotional support, social support, and patient empowerment are important components of health social networks, available both implicitly and explicitly. Implicitly, emotional support is experienced by seeing that there are others with similar conditions, that "I am not alone." Implicit emotional support is also felt by being a community member, participating in the process of creating a personal profile and recording health information, seeing how other non-medical professionals describe the same conditions and finding out what remedies others have tried. Emotional support is also offered explicitly in some health social networks. Site members may have the ability to comment on forums, publicly or privately message each other, give each other advice and transmit lightweight social greetings, such as hugs, as shown in excerpts from DailyStrength's activity feed.

In health social networks, people use technology to share their experiences with other individuals in the community through participatory approach evolved in cultural phenomena. According to Jenkins et al. (2006) a participatory culture is a culture with relatively low barriers to artistic expression and civic engagement, strong support for creating and sharing one's creations, and some type of informal mentorship whereby what is known by the most experienced is passed along to novices. Jenkins et al. (2006) define four forms of participatory culture: affiliations, expressions, collaborative problem solving and circulations as shaping the flow of media. Potential benefits of these forms of participation contain opportunities for peer-to-peer learning, the diversification of cultural expression, creative expression, civic engagement and a more empowered conception of citizenship, which could be applied in healthcare domain.

Teenagers today are utilizing interactive capabilities of the Internet as they create and share their own media creations. In many cases, these teens are actively involved in "participatory cultures". According to a study from the Pew Internet & American Life project (2005) more than one-half of all teens have created media content, and roughly one third of teens that use the Internet have shared content they produced. Fully half of all teens and 57 % of teens that use the Internet could be considered 'Content Creators'. They create blogs, webpage, post original art-works, photography, stories, and videos on line, download music or remixed online content into their own creation. According to Jenkins et al. (2006) participatory culture shifts the focus of literacy from individual expression to community involvement. The new literacy involves almost all social skills that are developed through collaboration and networking. Participatory culture is emerging as the culture absorbs and responds to the explosion of new media technologies that make it possible for average consumers to archive, annotate, appropriate, and re-circulate media content in powerful new ways. As Jenkins et al. (2006) claims, interactivity is a property of technology, while participation is a property of culture. Rather than dealing with each technology in isolation, we would do better to take an ecological approach, thinking about the interrelationship among all of these different com-munication technologies, the cultural communities that grow up around them, and the activities they support.

Many authors have argued that new participatory cultures represent an ideal learning environments. Gee (2004) calls such informal learning cultures "affinity spaces," asking why people learn more, participate more actively, engage more deeply with popular culture than they do with the contents of their textbooks. Affinity spaces offer powerful opportunities for learning, as Gee (2004) argues, because they are sustained by common endeavors that bridge differences in age, class, race, gender, and educational level, and because people can participate in various ways according to their skills and interests, because they depend on peer-to-peer teaching with each participant constantly motivated to acquire new knowledge or refine their existing skills, and because they allow each participant to feel like an expert while tapping the expertise of others.

For example, Black (2005a, b) finds that the "beta-reading" provided by online fan communities helps contributors grow as writers, mastering not only the basic building blocks of sentence construction and narrative structure, but also pushing them to be close readers of the works that inspire them. Participants in the beta-reading process learn both by receiving feedback on their own work and by giving feedback to others, creating an ideal peer to- peer learning community.

Affinity spaces, informal learning culture, are distinct from formal educational systems in several ways. While formal education is often conservative, the informal learning within popular culture is often experimental. A 2005 report on The Future of Independent Media (Blau 2005) argued that this kind of grassroots creativity is an important engine of cultural transformation. The Pew study suggests something more: Young people who create and circulate their own media are more likely to respect the intellectual property rights of others because they feel a greater stake in the cultural economy. Although there are opponents that discuss adverse health consequences that adults may inadequately supervise and interact with children about the media they consume and produce; or concerns about the moral values and commercialization in much contemporary entertainment. Yet, the focus on negative effects of media consumption offers an incomplete picture. These accounts do not appropriately value the skills and knowledge young people are gaining through their involvement with new media, and as a consequence, they may mislead us about the roles teachers and parents should play in helping children learn and grow.

Merkel et al. (2004) suggests that in the context of community technologies the role of designer goes beyond that of eliciting project requirements to include finding ways to seed ownership. Dearden and Light (2008) note that one of the emerging roles for designers working with community platforms are the up-skilling of community members. Garfinkel (1967) developed a specific approach called etnomethodology, defined below. It was used in the study of the ways that every day people use to detect, attribute significance and classify the actions of others and their own. On this occasion, it used explicitly for the first time the analysis of speech in order to understand interpretations and behaviors of the actors. In fact, very often people accompany their actions with a series of comments. Detection of the dialogue between the actors enters and then become part of the tools of sociological research on the side of observations, interviews and documentary analysis.

Social technologies shape the social interaction and co-experience of the product or service used by the participant. For example Ito et al. (2005) state that studies of camera phones have shown that the phones themselves become objects around which participation occurs. The success of online communities is attributed to "object centered sociality" as described by Engeström (2005) that emerges around specific videos, photos and collections. Leivrouw (2006) defines "new media" as the participatory design in the context of social technologies, and Boyd (2007) states that social technologies can be characterized by greater social participation in mediated contexts. Increasingly commercial, government and non-profit organizations are embracing social technologies in these mediated contexts as a way to support what Cottam (2010) defines as "mass participation". Boyd (2009) and Shirky (2008) state that the ease with which we can now connect, communicate, produce, share, replicate, locate and distribute information has had, and continues to have, a profound impact on our social, cultural and technological practices.

In mobile computing, this type of participation presents an important new shift in mobile device usage—from communication tool to 'networked mobile personal measurement instrument'. These new 'instruments' enable entirely new participatory urban lifestyles and create novel mobile device usage models.

There are various different forms of participation used in mobile computing literature to make citizens to become proactive in their involvement. Locative Media Narratives is a participatory storytelling, where narrative formats can arise from making use of the geo-locative and high-quality media creation abilities of modern mobile devices. Guided mobile video production is a technique of using mobile media and customized templates to guide the creation of video content and narrative storytelling. Local Connected Media is an approach to use mobile devices and new media to better connect communities and neighborhoods in a localized and culturally relevant way. The critical question is how can we apply these forms of participation to design socially connected, participative and holistic healthcare systems.

2.5 Health Awareness as a Background Study

Design and technology help people transition from the lifestyle they have to the lifestyle they want by helping them to create awareness and change their behavior. There are case studies and research projects focusing on awareness during action, where these actions created by mobile technologies enable individual participation towards a healthier life. Rubin et al. (1986) state that evaluations of educational programs have indicated that patient knowledge about asthma increases, however patients demonstrate no significant changes in asthma related to behaviors or outcomes. Such poor correlation between knowledge and behavior according to Rich et al. (2002) has left healthcare community at a loss. This indicates developing awareness and changing behavior is beyond providing knowledge to patient.

In theoretical background of awareness and behavior change in health, Prochaska et al. (1992) describes The Transtheoretical Model, the stages through which an individual progresses to intentionally modify addictive or other problematic behaviors;

- Precontemplation—no intention to change in the foreseeable future.
- Contemplation—seriously considering changing, but has not committed to taking action.
- Preparation—intends to take action in the next month and has unsuccessfully taken action in the past year.
- Action—has performed the desired behavior consistently for less than six months.
- Maintenance—has consistently performed the desired behavior for six or more months

The Transtheoretical Model suggests that a persuasive technology that targets precontemplators might focus on education. For contemplators, the design might focus on techniques for overcoming barriers or rewards for performing the desired behavior. For preparation stagers, it might focus on rewarding behaviors, even when the behavior is not consistent and increasing awareness of patterns of the behavior to encourage consistency. For action stagers, the design might focus on keeping track of progress to maintain consistency and possibly incorporate elements of social influence. For maintainers, it might focus on coping strategies for problems encountered previously and helping the individual realizes how she is becoming "the kind of person one wanted to be" (Prochaska et al. 1992, p. 12). However, because individuals are likely to use the technology throughout everyday life, social psychologists focus on how individuals manage their daily behaviors as part of a larger social context.

Presentation of Self in Everyday Life (Goffman 1959) addresses the social interactions that individuals manage daily. And because the desire for lifestyle change often emerges when the individual recognizes a conflict between her current and an ideal state, Cognitive Dissonance Theory (Festinger 1957) describes what happens when an individual realizes that her behaviors and attitudes are inconsistent.

The application of Presentation of Self and Cognitive Dissonance Theory to persuasive technology is novel, however use of the theories in research and practice is not. For example, Aoki and Woodruff (2005) use Presentation of Self to analyze online community interactions and develop collaborative systems.

Presentation of Self in Everyday Life describes how individuals attempt to manage the impressions they want others to have of them. This impression management is a constant process. Presentation of Self uses the metaphor of the theatre stage to describe how people interact with others. The performance encompasses "all the activity of an individual which occurs during a period marked by his continuous presence before a particular set of observers" (Goffman 1959, p. 22). The individual performs for an audience. The audience consists of those who observe the performance. The individual and audience are the participants in the performance. Non-participants are outsiders. The individual has a personal front, which consists of traits such as gender, age, size, looks, and clothing.

As Goffman (1959) states, a given performance has two regions: Front and backstage. Front stage is where the individual knowingly performs. Backstage is "a place, relative to a given performance, where the impression fostered by the performance is knowingly contradicted as a matter of course...Here the performer can relax; he can drop his front, forgo speaking his lines, and step out of character...[it is] where the performer can reliably expect that no member of the audience will intrude" (Goffman 1959, p. 112.) Impression management describes how the individual moves between front and backstage and controls access to backstage.

Other important concepts include dramatic realization, misrepresentation, and secret consumption. Dramatic realization occurs when the individual draws attention to facts that may go unnoticed. Misrepresentation suggests that individuals may be incented to misrepresent facts. Secret consumption explains that, "If an individual is to give expression to ideal standards during his performance, then he will have to forgo or conceal action which is inconsistent with these standards. When this inappropriate conduct is itself satisfying...then one commonly finds it indulged in secretly; in this way the performer is able to forgo his cake and eat it too" (Goffman 1959, p. 41.) Secret consumption often results from idealization—the idealized impression the individual may attempt to convey of her behaviors.

According to Self in Everyday Life Theory, technology to encourage lifestyle behavior change must support fundamental impression management needs. Effective support implies that the individual should be in control of information about herself that is collected and how that information is used. For example, it may be important to provide the ability for the individual to disguise something about her activities. The technology may also need to enable the individual to misrepresent something about her behavior, perhaps to support secret consumption. It is often the case that a technology that seeks to represent "perfect information" may not effectively support an individual's basic need to control backstage access, yet it is important to give individuals control over their backstage. The technology needs to enable the individual to perform differently for different audiences. If the technology allows an audience member or outsider to access her backstage, the individual may perceive a violation of her privacy, which could result in her abandoning the technology.

Cognitive Dissonance Theory (Festinger 1957) explains what happens when an individual realizes that her attitudes and behaviors are inconsistent. When that happens, the individual will experience psychological discomfort (or dissonance). Because this dissonance is psychologically uncomfortable, the individual will be motivated to reduce or eliminate the dissonance. Her motivation depends on how important the beliefs or behaviors are to her. That is, the more important they are to the individual, the more likely she will try to reduce or eliminate the dissonance.

When an individual is motivated to reduce or eliminate dissonance, she can change her behavior or knowledge. She may also reduce the importance of the dissonance by actively learning about other things that are more harmful than whatever is causing the dissonance or by actively avoiding information or situations that may produce (or reinforce) the dissonance. For example, a smoker may experience cognitive dissonance because of the health risks she believes smoking causes. She can change her behavior—stop smoking, change her knowledge—stop believing that health risks are involved with smoking, rationalize the dissonance— focus on how she does not behave in ways that she can convince herself are more harmful than smoking, perhaps binge eating, or avoid information and situations that reinforce the dissonance—avoid warnings of smoking's health risks.

Cognitive Dissonance Theory suggests that a persuasive technology to encourage lifestyle behavior change should address whichever factors may prevent the individual from incorporating the change into her everyday life (i.e., by helping her change her behavior to match her attitudes). For example, the technology should help the individual remain focused on her commitment to change and her relevant patterns of behavior. Supporting reflection, or one's ability to reach conclusions relying on one's memory or personal experience is one of the most persistent yet exclusive goals. While general usefulness of health monitoring applications is undoubted, many of the evaluation studies carried on so far focused on qualitative assessment and participants' self-reports (Consolvo et al. 2009). The awareness provided by the technology should be persistently available and easy to access, yet subtle enough so as to support occasional needs for information and situation avoidance. There are several methods to develop awareness by self-documentation tools such as diary methods.

2.5.1 The Use of Mobile Diary Studies in Healthcare Context

2.5.1.1 The Act of Self-Reporting

There are several studies published on the topic of capturing life experiences of an individual, such as wearable camera Sensecam (Gemmell et al. 2004); online media weblog (Kelliher 2004); group story sharing (Appan et al. 2004); life logging

(Sellen et al. 2007); capture and access (Abowd and Mynatt 2000). All these applications, which are self-reporting methods, have their own limitations in terms of collecting data, processing information, and retrieving experiences from individuals.

In self-reporting, participants are responsible for the data collection, gathering contextual data over-time and in situ, without the physical presence of researchers. Self-reporting can provide access into the private, personal and environmental aspects of people's lives that are often difficult, or impossible, to access through traditional methods such as observations or interviews. The personal reflection inherent in self-reporting makes aspects available that would otherwise remain tacit. Since so much of individuals' lives are routinized and automatic, it is not until they are asked to document or consider certain activities that they are able to identify key junctures in our own understanding of a behavior.

Self-reporting studies can take many different forms and the degree of formal structure is one of the things that differentiates approaches and determines the type of material collected. For example, in the Electronic Sampling Method approach known as ESM (Larson and Csikszentmihalyi 1983), the participant is directed to systematically log specific things at specific times. In more open-ended approaches such as the ones reported, data collection is only semi-structured around a particular topic (Gaver et al. 1999; Hulkko et al. 2004; Masten and Plowman 2003; Palen and Salzman 2002; Sanders 2008). In this case participants are treated as active con-tributors and interpreters in the design process and select what, how and when to report. This encourages more playful and creative representations, important to an explorative and collaborative approach.

2.5.1.2 Mobile Diaries as Self-Reporting Method

Over the last ten years digital, online and mobile technologies have been incor-porated into self-reporting methods in a range of ways. These everyday tools can be easily integrated into people's daily lives and support the generation of a range of different media forms such as video, images, text and audio.

Mobile diaries are one of the life capturing tools that has been studied around the implementation of self-documentation methods in healthcare. These mobile diaries challenge new methods for self-reporting and new applications to enrich and integrate participatory design methods. Mobile diaries are a hybrid method that incorporates many of the creative and playful aspects of probes. A range of different analog and digital technologies are used that allow participants to share and reflect on various dimensions of their day-to-day life. Mobile diaries can be used as a fieldwork method to discover and extract knowledge in the early stages of design process to immerse into people's everyday life. This exploratory approach to self-reporting allows participants to create and share their everyday life.

The self-reporting through mobile diaries could contain different strategies such as writing text, or audio and video recording- consultation diaries. Generally, over the period of the study participants receive prompts, questions and reminders through their mobile phone and blog. Through these incentives, they create collages, videos and blog messages, and send reports, which appear on the blog where other participants could see and reflect on their behaviors.

The output of the study (Hagen and Rowland 2010) is a particularly provocative, experiential and sensorial insight into participant's lives with people's own words in their day-to-day lives. Using their phone, participants capture images, text and audio and send to the blog throughout the day. Through these reports researcher can track events, locations, and participant's emotions across the days and weeks. Over time, daily rhythms and habits emerge. As Masten and Plowman (2003) state, real-time reporting increases the sense of immersion in people's lives as we experience the activities 'as they happen'. This is complemented by more reflective accounts viewing the diaries via the blog or on the video camera. The use of video encourages in-depth descriptive accounts of events and surroundings from the participant's perspective.

In addition to descriptions of events and activities, a mobile diary also reports capturing emotions, feelings and inner thoughts. The example above shows emotional reactions and descriptions of personal feelings at particular moments through words or metaphors or symbols. The process of self-reporting is an intervention designed to allow people to self-reflect and share aspects of their daily life. This process can also trigger participants to question their choices and everyday behaviors as discussed by Grinter and Eldridge (2003). These activities help to reveal emotional and infrastructural barriers to behavioral changes.

Materials such as emotions, information, behavioral data generated from mobile diaries can be used in numerous ways. For designers, the visual nature of the material allows for more active interpretation than written research reports. For participants, the process of doing the mobile diaries means they are better equipped to reflect on and analyze their own practices. This also allows participants to become active interpreters of the material and what it might mean for future designs during follow up interviews and workshops.

The material generated through mobile diaries should not be reduced down into a traditional written report. As Mattelmäki and Battarbee (2002) state, the raw form of the material and the subjective picture it provides of the participants' lives and world-view not only provides additional information about them but also is essential to create connection with participants. However, this needs to be balanced with normative expectations of a 'research outcome', and the data transmitted to other healthcare stakeholders in a meaningful way. The development of a multimedia report should support all the different formats of material generated. As Hagen and Rowland (2010) found out in their study, the results of the reports introduce the participants' lifestyle through their own words and images, illuminate the themes that have emerged and identify some future possibilities to be considered. There is also a significant value in creating opportunities for co-interpretation of the material by practitioners and other actors of the project.

Video diaries in healthcare, unless paper diaries, provide a more "direct" understanding of participants' experiences than is afforded by data that are "controlled" by the researcher. As Gibson (2005) has articulated in her argument, how participants construct their video accounts provides a valuable source of analyzable data. As in the study of asthma by Rich et al. (2000) Video Intervention Assessment was developed to determine whether medical information gathering might be augmented by video diaries created by patients to show clinicians the realities of managing chronic disease in the context of their lives. The unstructured user-generated data is being analyzed by software by structuring in the post-production.

Psychologists Csikszentmihalyi and Larson (1987) in their study state that they have recognized the need to monitor patients in situ versus asking a participant to recall an event a week or more after the fact. Previous studies show that many factors affect the memory of participants recalling past experiences. Recall can be caused by (i) the current emotional state of the participant, (ii) the length of time the participant is asked to recall, (iii) the participant sitting in a foreign environment. Questioning the results of studies rely solely on participant's long-term memory. Instead of trying to account all of these, researchers can alleviate the issue by having participants' record events as they happen or immediately after.

Adding diary functionality to mobile phones has advanced healthcare studies in managing chronic diseases and healthcare prevention. Mobile applications and social platforms allow individuals to express their thoughts and feelings about their healthcare treatment or prevention process. There are health-related diaries such as PalmOS Headache Diary, a project developed by Holroyd and Chen (2000) that allows one to record migraine activities, associated migraine symptoms, migraine related disability, and medication use. Headache Hero, Headache Diary, iHeadache, MiSelf are some migraine management applications that track symptoms, duration, severity, impact, triggers and medications. Headache Relief Diary is an application developed by New York Headache Center, in which the diary consists of weather, food, sleep cycle, and hormonal changes and offers physical treatment methods. Diary as a personal narrative form in a social platform comes into the scene as Patient Voices. In this platform patients use video as a means of communication with their emotions, feelings and ideas. They express themselves by going through a chronic disease experience, their success stories recovering from a disease where they get social support. Storytelling helps to improve the phase of the cure of the disease.

Diary-like self-documentation such as video consultation could be used for early diagnosis and follow-up care of chronic conditions such as ulcer, migraine, diabetes, for advice and instructions regarding assistive technology or as remote support for handling technical failures with artificial healthcare equipment, thus contributing to both quality and efficiency of care. Klammer et al. (2011) developed a social media- based collaborative platform, where participants created photos, short videos or text about daily work situations. The tool itself served to collect data but also created collaborative awareness. Preliminary research results, evaluated through interview questions, showed that participants wanted additional structure and guidance in documenting, and more specific focus on what to look for at work. Not knowing what exactly to show on video put too much pressure on the participants. The problem was not a lack of time or motivation, but related to the documentation technique itself. Another important result reported from this study was the longer the participant involves in the all design phase; the stronger relations build up among stakeholders. This allows participants to better understand how to participate and how to input their influences in the process.

Another example can be found in the study by Mahmud et al. (2008) where they developed a home-based storytelling support system for Aphasics, primarily intended to be used in their post-rehabilitation period. The focus of the study was on the creation of daily stories with the help of passively captured materials, hence emphasizing the need for fairly effortless interaction from the side of the end user. The premise of their project was to enable Aphasics to share their daily stories that would help them to become more active socially and to reengage in their preferred life style. The project included passive capturing with a webcam application with a motion detection utility, which usually generates a huge collection of images of ongoing activities at the place. The camera used a pre-programmed algorithm, which filtered low-quality images. However, still there were images with no person in the photo, therefore filtering the photos to feed the story creation was a main challenge. Therefore, the partner had provided subjects only content rich images for a story creation.

References

Abowd, G. D., & Mynatt, E. D. (2000). Charting past, present and future research in ubiquitous computing. *ACM Transactions on Computer-Human Interaction, 7*(1), 29–58.

Akyildiz, I., & Rudin, H. (2001). Pervasive computing. *Computer Networks The International Journal of Computer and Telecommunications Networking, 33*(4), 371.

Aoki, P. M., & Woodruff, A. (2005). Making space for stories: ambiguity in the design of personal communication systems. *Proceedings of CHI 2005, Portland, OR, USA* (pp. 181–90).

Appan, P., Sundaram, H., & Birchfield, D. (2004). Communicating everyday experiences. In *Proceedings of the 1st ACM Workshop on Story Representation, Mechanism and Context, SRMC '04. ACM, New York, NY* (pp. 17–24).

Barkhuus, L. (2004). Privacy in location-based services, concern vs. coolness. MobileHCI 2004 workshop: Location system privacy and control.

Benkler, Y. (2006). The wealth of networks: How social production transforms markets and freedom (p. 3). New Haven: Yale University Press. ISBN 0-300-11056-1.

Black, R. W. (2005a). Access and affiliation: The literacy and composition practices of english language learners in an online fanfiction community. *Journal of Adolescent and Adult Literacy, 49*(2), 118–128.

Black, R. W. (2005b). Online fanfiction: What technology and popular culture can teach us about writing and literacy instruction. *New Horizons for Learning Online Journal, 11*(2).

Blau, A. (2005). The future of independent media. *Deeper News, 10*(1). Available online at www.gbn.com/ArticleDisplayServlet.srv?aid=34045

Boyd, D. (2007). Social network sites: Public, private, or what?

Boyd, D. (2009). Taken out of context: American teen sociality in networked publics. PhD Thesis. University of California, Berkeley.

Budinger, T. F. (2003). Bio monitoring with wireless communications. *Annual Review of Biomedical Engineering, 5*, 383–412.

Castells, M. (1996). The rise of the network society. *The information age: Economy, society and culture* (Vol. 1). London: Blackwell

Centola, D. (2010). The spread of behavior in an online social network experiment. *Science, 329,* 1194–1197.

Christakis, N. A., & Allison, P. D. (2006). Mortality after the hospitalization of a spouse. *The New England Journal of Medicine, 354,* 719–730.

Christakis, N., & Fowler, J. H. (2007). The Spread of Obesity in a Large Social Network over 32 Years. *The New England Journal of Medicine, 357,* 370–379. doi: 10.1056/NEJMsa066082.

Contractor, N., & Bishop, A. P. (2000). Reconfiguring community networks. The case of prairie-know. In T. Ishida & K. Isbister (Eds.), *Digital cities, technologies, experiences and future perspectives.* Berlin, Heidelberg: Springer.

Consolvo, S., McDonald, D., & Landay, J. L. (2009). Theory-driven design strategies for technologies that support behavior change in everyday life. Conference on Human Factors in Computing Systems. *Proceedings of the 27th international conference on Human factors in computing systems.*

Cottam, H. (2010). Participatory systems: Moving beyond 20th century institutions. *Harvard International Review, 31*(4), 50–55.

Crespo, R. (2007). Virtual community health promotion. *Preventing Chronic Disease, 4*(3). A75. PMID: 17572979.

Crow, G., & Allan, G. (1994). *Community life. An introduction to local social relations.* Hemel Hempstead: Harvester Wheatsheaf.

Csikszentmihalyi, M., & Larson, R. (1987). Validity and reliability of the experience-sampling method. *The Journal of Nervous and Mental Disease, 175*(9).

Dearden, A., & Light, A. (2008). Designing for e-Social Action, An Application Taxonomy Design Research Society (DRS).

Engeström, J. (2005). Why some social network services work and others don't—Or: The case for object-centered sociality. Retrieved from http://www.zengestrom.com/blog/2005/04/why_some_social.html, August 2008.

Eysenbach, G. (2001). What is ehealth. *Journal of Medical Internet Research.*

Feiner, S. K. (2000). Environment management for hybrid user interfaces. *IEEE Personal Communications, 7,* 50–53.

Festinger, L. (1957). *A theory of cognitive dissonance.* Stanford, CA: Stanford University Press.

Frost, J. H., Massagli, M. P., Wicks, P., & Heywood, J. (2008). How the social web supports patient experimentation with a new therapy: The demand for patient-controlled and patient-centered informatics. *AMIA Annual Symposium Proceedings* (Vol. 6, pp. 217–21).

Fogg, B. J. (2003). *Persuasive technology: Using computers to change what we think and do.* San Francisco, CA: Morgan Kaufmann Publishers.

Fox, S. (2009). The social life of health information. June 11, 2009. Retrieved from: http://www.pewinternet.org/Reports/2009/8-The-Social-Life-of-Health-Information.aspx

Fox, S., & Duggan, M. (2012). Mobile Health 2012. Pew Research Center. Accessed at http://www.pewinternet.org/files/oldmedia//Files/Reports/2012/PIP_MobileHealth2012_FINAL.pdf

Garfinkel, H. (1967). *Studies in ethnomethodology.* Englewood Clffs, NJ: Prentice Hall.

Gaver, B., Dunne, T., & Pacenti, E. (1999). Design: Cultural probes. *Interactions, 6,* 21–29.

Gee, J. P. (2004). *Situated language and learning: A critique of traditional schooling.* New York: Routledge.

Gemmell, J., Williams, L., Wood, K., Lueder, R., & Bell, G. (2004). Passive capture and ensuing issues for a personal lifetime store. In *Proceedings of the 1st ACM Workshop on Continuous Archival and Retrieval of Personal Experiences, CARPE'04* (pp. 48–55). New York, NY: ACM.

Gibson E. B. (2005). Co-producing video-diaries: The presence of the "Absent" researcher. *International Journal of Qualitative Methods, 4.*

Giustini, D. (2006). How web 2.0 is changing medicine: Editorial. *British Medical Journal, 333,* 1283–1284.

Godoe, H. (2000). Innovation regimes, R&D and radical innovations in telecommunications. *29* (9), 1033–1046. doi:10.1016/S0048-7333(99)00051-7

Goffman, E. (1959). *The presentation of self in everyday life.* New York, NY, USA: Doubleday Anchor.

Grayson, A.C.R., Voskerician, G., Lynn, A., Anderson, J. M., Cima, M. J., & Langer, R. (2004). Differential degradation rates in vivo and in vitro of biocompatible Poly(Lactic Acid) and Poly (Glycolic Acid) Homo- and Co-Polymers for a Polymeric Drug-Delivery Microchi. *Journal of Biomaterials Science, Polymer Edition, 15,* 1281–1304.

Grinter, R., & Eldridge, M. (2003). Wan2tlk? Everyday text messaging. *Proceedings of the CHI 2003* (pp. 441–448).

Hakkila, J., & Chatfield, C. (2005). It is like if you opened someone else's letter: user perceived privacy and social practices with SMS communication. In *Proceedings of the 7th International Conference on Human Computer Interaction with Mobile Devices and Services* (pp. 219–222). doi: 10.1145/1085777.1085814.

Hagen, P., & Rowland, N., (2010). *Mobile diaries: Discovering daily life.* Johnny Holland (Online Magazine).

Hawkes, N. (2005). *More people consult Google over health.* Times Online. June 6, 2005.

Hemment, D. (2006). Locative arts. *Leonardo, 39*(4), 348–355.

Hill, C. T. (2007). The post-scientific society. *Issues in Science and Technology.* Fall.

Holroyd, K. A., & Chen, Y. (2000). *A hand-held computer headache diary program: Monitoring headaches, medication use, and disability in real time.* Paper presented at the American Headache Society Convention, Quebec, Canada.

Howe, J. (2008). Crowdsourcing: Why the power of the crowd is driving the future of business. NY, USA: Crown Publishing Group New York.

Hughes, B., Joshi, I., & Wareham, J. (2008). Health 2.0 and medicine 2.0: Tensions and Controversies in the Field. *Journal of Medical Internet Research, 10*(3), e23. (Adapted from Jane Sarasohn-Kahn's "Wisdom of Patients" report, by Matthew Holt, Last updated June 6, 2008).

Hulkko, S., Mattelmäki, T., Virtanen, K., & Keinonen, T. (2004). Mobile probes. *NordiCHI 04,* Tampere, Finland.

Istepanian, R., & Lacal, J. (2003). Emerging mobile communication technologies for health: Some imperative notes on m-health. In *Engineering in Medicine and Biology Society the 25th Silver Anniversary International Conference of the IEEE.* Cancun Mexico: IEEE.

Istepanian, R. S. H., & Wang, H. (2003). Telemedicine in UK, in European TLEMEDICINE glossary concepts, standards, technologies and users. In Beolchi, L. (Ed), *European commission information society directorate 5th edition,* (pp. 1159–1165).

Ito, M., Okabe, D., & Matsuda, M. (Eds.). (2005). *Personal, portable and pedestrian.* Cambridge: MIT.

Jégou, F., & Manzini, E., (2008). *Collaborative services. Social innovation and design for sustainability.* Milan: Poli.design edizioni.

Jenkins, H., Clinton, K., Purushotma, R., Robinson, A., & Weigel, M. (2006). *Confronting the challenges of participatory culture: Media education for the 21st century.* MacArthur Foundation.

Kahn, J. G., Yang, J., & Kahn, J. S., (2008). *The relationship among economic development, health, and the potential roles of mhealth.* The conference material of 'making the eHealth connection'.

Kelliher, A. (2004). Everyday cinema. In *Proceedings of the 1st ACM Workshop on Story Representation, Mechanism and Context, SRMC'04,* (pp. 59–62). New York: ACM.

Kindberg, T., Spasojevic, M., Fleck, R., & Sellen, A., (2005). The Ubiquitous camera: An in-depth study of camera phone use. In: IEEE Pervasive Computing. IEEE Educational Activities Department, Piscataway, NJ, USA.

Klammer, J., van den Anker, F., & Janneck, M. (2011) *Participatory service innovation in healthcare: the case of video consultation for paraplegics.* Participatory innovation conference (pp. 290–297), January 13–15th, 2011, Sønderborg, Denmark.

Langheinrich, M. (2009). Privacy in ubiquitous computing. In Krumm, J. (Ed.), *Ubiquitous computing*. CRC Press.

Larson, R., & Csikszentmihalyi, M. (1983). The experience sampling method. In H. Reis (Ed.), *Naturalistic approaches to studying social interaction: New directions for methodology of social and behavioral science*. San Francisco: Jossey-Bass.

Laxminarayan, S., & Istepanian, R. S. H. (2000). Unwired e-medicine: The next generation of wireless and internet telemedicine systems. *IEEE Transactions on Technology in Biomedicine, 4*(3), 189–193.

Lenert, L., & Kaplan, R. M. (2000). Validity and interpretation of preference-based measures of health-related quality of life. *Medical Care, 2000*(38), 138–150.

Lievrouw, L. A. (2006). *Oppositional and activist new media: remediation, reconfiguration, participation*. PDC.

Lin, J. L, Mamykina, L., Lindtner, S., Delajoux G., & Strub, H. B. (2006) *Proceedings of the UbiComp 2006. Fish'n'Steps: Encouraging physical activity with an interactive computer game* (pp. 261–278). Springer.

Liu, L. S., Hirano, S. H., Tentori, M., Cheng, K. G., George, S., Park, S. Y. (2011). Improving communication and social support for caregivers of high-risk infants through mobile technologies. In *Proceedings of the CSCW 2011* (pp. 475–484).

Mahmuh, A. A., Aliakseyeu, D., & Martens, J.B. (2008). Enabling storytelling by Aphasics in an augmented home environment. In *BCS-HCI '08 Proceedings of the 22nd British HCI Group Annual Conference on People and Computers: Culture, Creativity, Interaction* (Vol. 2). ISBN: 978-1-906124-06-9.

Mattelmäki, T., & Battarbee, K. (2002). *Empathy Probes Paper presented at the PDC 2002*, Malmö, Sweden.

Masten, D., & Plowman, T. (2003). Digital ethnography: The next wave in understanding the consumer experience. *Design Management Journal, 14*(2), 75–81.

Merkel, C. B., Xiao, L., Farooq, U., Ganoe, C. H., Lee, R., & Carroll, J. M. (2004). *Participatory design in community computing contexts: Tales from the field proc*. PDC.

Moraveji, N., Oppezzo, M., Habif, S., & Pea, R. A (2011). Theoretical model of calming technology: Designing to mitigate stress and increase calm. In *Proc. of Workshop on Interactive Systems in Healthcare 2011*.

Olla, P. (2005). Evolution of GSM network technology. In M. Pagani (Ed.), *Encyclopedia of multimedia technology and networking* (pp. 290–294). Hershey, PA: Idea Group Reference.

Palen, L., & Salzman, M. (2002). *Voice-mail diary studies for naturalistic data capture under mobile conditions*. Louisiana, USA: CSCW.

Peltonen, P., Salovaara, A., Jacucci, G., Ilmonen, T., Ardito, C., & Saarikko, P. et al. (2007). Extending large scale event participation with user created mobile media on a public display. In *Proceedings of the MUM 2007*.

Prochaska, J., DiClemente, C., & Norcross, J. (1992). In search of how people change: Applications to addictive behaviors. *American Psychologist, 47*(9), 1002–1114.

Prochaska, J. O., & Velicer, W. (1997). The transtheoretical model of health behaviour change. *American Journal of Health Promotion, 12*(1), 38–48. doi:10.4278/0890-1171-12.1.38

Puikkonen, A., Venta, L., Beekhuyzen, J., & Hakkila, J. (2008). Playing, performing, reporting- a case study of mobile minimovies composed by teenage girls. OZCHI.

Rheingold, H., (2007). Using Participatory Media and Public Voice to Encourage Civic Engagement. In John D., & Catherine T (Eds.), *MacArthur foundation series on digital media and learning* (pp. 97–118).

Rich, M., Lamola, S., Amory, C., & Schneider, L. (2000). Asthma in life context: Video Intervention/prevention assessment, pediatrics., division of adolescent/young adult medicine. *Childrens Hospital, 105*(3Pt 1), 469–77. (Boston, MA).

Rich, M., Patashnick, J., & Chalfen, R. (2002). Visual illness narratives of asthma: Explanatory models and health-related behavior. *American Journal of Health Behavior, 26*(6), 442–453.

Rubin, D. H., Leventhal, J. M., Sadock, R. T., Letovsky, E., Schottland, P., & Clemente, I. (1986). Educational intervention by computer in childhood asthma: A randomized clinical trial testing the use of a new teaching intervention in childhood asthma. *Pediatrics, 11*, 1–77.

Rutter, J. (2001). From the sociology of trust towards a sociology of 'e-trust'. *International Journal of New Product Development and Innovation Management, 2*(4), 371–385.

Sanders, L. (2008). An evolving map of design practice and design research, Interactions (November–December, pp. 13–17).

Savidis, A., & Stephanidis, C. (2005). Distributed interface bits: Dynamic dialogue composition from ambient computing resources. *Personal and Ubiquitous Computing, 9*(3), 142–168.

Sellen A., Fogg A., Aitken M., Hodges S., Rother C., & Wood K. (2007). Do lifelogging technologies support memory for the past? In *Proc. of CHI 2007* (pp. 81–90), New York, ACM.

Shirky, C. (2008). *Here comes everybody.* London: Penguin Press.

Smith, A. (2010). Mobile Access Report 2010, Pew Internet and American Life Project, October 19, 2010. http://www.pewinternet.org/Reports/2010/Mobile-Access-2010.aspx

Smith, K. P., & Christakis, N. A. (2008). Social networks and health. *Annual Review of Sociology, 34*, 405–429.

Stachura, M., & Khasanshina, E. (2007). Telehomecare and remote monitoring: An outcomes overview.

Swan, M. (2009). Emerging patient-driven health care models: An examination of health social networks, consumer personalized medicine and quantified self-tracking. *International Journal of Environmental Research and Public Health, 6*(2).

Tan, H., & Ng, J. H. K. (2006). Googling for a diagnosis—use of Google as a diagnostic aid: internet based study. *BMJ, 333*, 1143–1145.

Tapscott, D., & Williams A. D. (2007). *Wikinomics: How mass collaboration changes everything* (320 pp). New York: Penguin. (Int. Journal of Communication).

Taylor, A. S., & Harper, R. (2003). The gift of the gab?: A design oriented sociology of young people's use of mobiles. *Computer Supported Cooperative Work, 12*(3), 267–296.

Tsai, C. C., Lee, G., Raab, F., Norman, G. J., Sohn, T., Griswold, W., et al. (2007). Usability and feasability of PmEB: A mobile phone application for monitoring real time caloric balance. *Mobile Networks and Applications, 12*(2–3), 173–184.

Trogemann, G., & Pelt, M. (2006). Citizen media. Technological and social challenges of user driven media BB Europe.

Wagner, D., Lopez, M., Doria, A., Pavlyshak, I., Kostakos, V., Oakley, I., & Spiliotopoulos, T. (2010). Hide and seek: Location sharing practices with social media. *MobileHCI, 2010*, 55–58.

Wellman, B. (2000). Physical place and cyber-place: The rise of networked individualism. *Paper presented to Community Informatics: Connecting communities through the web.* University of Teeside. April 26–28th, 2000.

Wellman, B. (2001). The rise of networked individualism. In Leigh Keeble (Ed.), *Community Networks Online.* London: Taylor & Francis.

Wilmott, P. (1986). *Social networks, informal care and public policy.* Policy Studies Institute: London.

Chapter 3
Research Strategies for Mobile Healthcare

3.1 Practiced-Based Approach

Practice-based research comprehends various approaches, models, and methods aimed at applying practical considerations to theoretical knowledge. Practice is comprised of research findings derived from the systematic collection of data through observation, experiment and analysis of results.

This book promotes the idea of research through design based upon a generic structure of learning and designing, which has been derived from practice. The proposed paradigm of design research is the generic design process and not the scientific process that guides design research. As Jonas (2005) states, research through design covers the whole process development and design as an institution for human-centered innovation and supporting design as a discipline. Research is guided through design process logic, and design is supported and driven by phases of scientific research and inquiry. Schön (1983) introduces the idea of design as a reflective practice where designers reflect back on the actions taken in order to improve design methodology.

It is not science as a method, but science as a guiding paradigm for design, which is being called into question. Examining design as processes reveals more clearly what is impossible and enables us to identify reliable knowledge. This view adopts the circular and reflective "trial and error" models of generative world appropriation, as stated by Dewey (1986), Von Foerster (1981), Glanville (1982), Schön (1983), Swann (2002) and many others. Since there are debates in design research in scientific about research through design or design through research, Sanders (2011) argues the relation between research-led thinking and design-led thinking.

The map of "Design Research and Practice" (Sanders and Stappers 2008) illustrates two intersecting dimensions: approach and mind-set for design research and practice. Design research has traditionally been a research-led thinking

© The Author(s) 2016
P. Arslan, *Mobile Technologies as a Health Care Tool*,
PoliMI SpringerBriefs, DOI 10.1007/978-3-319-05918-1_3

approach that has a long history driven by social scientists and engineers. The design-led thinking approach is a recent way of thinking and is mostly driven by design practitioners. In the practice of design research, there are two opposing mindsets: the experts' mindset in which people were referred to as "subjects", "users", or "consumers", and a participatory mindset that refers to people as 'true experts in domains of experience such as living, learning, working and co-creating in the design process with co-workers'.

The user-centered perspective uses research-led approaches coming primarily from marketing and social sciences to make incremental improvements to existing products and services. The design-led perspective uses design thinking and has potential end-users as being participants in the early front end of the process. The co-creation perspective puts the tools and methods of design thinking into the hands of people who will be future end-users and other stakeholders early in the front end of the product development process. As Jefferys and Lashof (1991) stated, practice-based research gives balanced consideration to the co-equal responsibilities of research: (i) to educate practitioners, and (ii) to generate new knowledge of science and practice. A substantial part of all research is aimed at supporting and improving some kind of design practice. The goal is to create methods, tools, techniques, and approaches that can be used by design practitioners to improve their design ability and the quality of design outcome. However, there is a thin line between developing tools for design practice and for design research. The researcher judges the tool based on how well it is a result of good research, while the practitioner judges the practical usefulness of design.

3.2   Action Research Strategy

Action research, as defined by Reason and Bradbury (2001), is a participatory process concerned with developing practical knowledge in the pursuit of human purposes. This strategy seeks to bring together action and reflection, theory and practice, in participation with others, in the pursuit of practical solutions to issues of pressing concern to people, and more generally the flourishing of individual persons and their communities. O'Brien (2001) defines action research as "learning by doing"—a group of people identify a problem, do something to resolve it, see how successful their efforts were, and if not satisfied, try again. It is an iterative process with cycles, where the process has been reconsidered for an improvement of the previous cycles to reach the outcome of the research. Since the numbers of cycles are changing in the process depending on the research approach and setting, there are two models proposed: Simple Action Research Model and Detailed Action Research Model.

Kemmis and McTaggart (1988) have developed a simple model of the cyclical nature of the typical action research process where each cycle has four steps: plan, act, observe, and reflect. Whereas Susman (1983) distinguishes five phases to be conducted within each research cycle. In both models, initially a problem is identified and data is collected for a more detailed diagnosis. This is followed by a collective postulation of several possible solutions, from which a single plan of action emerges and is implemented. Data on the results of the intervention are collected and analyzed, and the findings are interpreted in light of how successful the action has been. At this point, the problem is reassessed and the process begins another cycle. This process continues until the problem is resolved.

According to the definition of Gilmore et al. (1986), "Action research...aims to contribute to the practical concerns of people in an immediate problematic situation and to further the goals of social science simultaneously. Thus, there is a dual commitment in action research to study a system and concurrently to collaborate with members of the system in changing it in what is together regarded as a desirable direction. Accomplishing this twin goal requires the active collaboration of researcher and stakeholders, and thus it stresses the importance of co-learning as a primary aspect of the research process." But action research goes beyond the notion that theory can inform practice, to a recognition that theory can and should be generated through practice. This approach is supported by the notion of only being useful so far as it is put in the service of a practice focused on achieving positive social change.

As Brydon-Miller et al. (2003) described, both practice and theory can benefit from combining action and research. Action research is regarded as research that is normally carried out by practitioners. It enables the researcher to investigate a specific problem that exists in practice. According to Landman (1988: 51) this requires that the researcher should be involved in the actions that take place. A further refinement of this type of research according to Jacobs et al. (1992: 431) is that the results obtained from the research should be relevant to the practice. In other words, it should be applicable immediately. This means that the researcher, as an expert, and the practitioner jointly decide on the formulation of research procedures, allowing the problem to be solved.

An action researcher differs from a regular consultant or a general professional practitioner by studying the problem systematically and ensuring that the intervention is informed by theoretical considerations. The role of action researcher is to implement the action research method in such a manner as to produce a mutually agreeable outcome for all participants, with the process being maintained by them afterwards. To accomplish this, it may necessitate the adoption of many different roles at various stages of the process, including those of planner leader, catalyzer, teacher, listener, synthesizer, facilitator, designer, observer and reporter. The main role, however, is to nurture local leaders to the point where they can take responsibility for the process. This point is reached when they understand the methods and is able to carry on the process when the initiating researcher leaves. The researcher's role is primarily to take time to facilitate dialogue and foster reflective analysis among the participants, provide them with periodic reports, and write a final report when the researcher's involvement has ended.

Action research is more a holistic approach to problem solving, rather than a single method for collecting and analyzing data. Thus, it allows using several different research tools during the project. These various methods, which are generally common to the qualitative research paradigm, include: Keeping a research journal, data collection and analysis, participant observation recordings, question-naire surveys, structured and unstructured interviews, and case studies. Action research's primary goal is to focus on turning people involved into researchers. People are more willing to apply what they have learned, when they do it them-selves. It has a social dimension when the research takes place in real-world situ-ations, and aims to solve real problems. Since, the initiating researcher, unlike in other disciplines, makes no attempt to remain objective, it openly acknowledges his/her bias to the other participants.

Action research is characterized by the following four features (Jacobs et al. 1992):

- Problem-aimed research focuses on a special situation in practice. Seen in a research context, action research is aimed at a specific problem recognizable in practice, and of which the outcome problem solving is immediately applicable in practice.
- Collective participation. A second characteristic is that all participants (in par-ticular researchers and practitioners) form an integral part of action research with the exclusive aim to solve the identified problem.
- Type of empirical research. Thirdly, action research is characterized as a means to change the practice while the research is going on.
- Outcomes of research cannot be generalized. Lastly, action research is charac-terized by the fact that problem solving, seen as renewed corrective actions, cannot be generalized, because it needs to comply with the criteria set for a scientific character.

Action research is mostly used in real situations, rather than in contrived, experimental studies, since its primary focus is on solving real problems. It is often the case that those who apply this approach are practitioners who wish to improve understanding of their practice, social change activists trying to mount an action campaign, or, more likely, academics who have been invited into an organization by decision-makers aware of a problem requiring action research, but lacking the requisite methodological knowledge to deal with it. Because action research is carried out in real-world circumstances, and involves close and open communica-tion among the people involved, as Winter (1996) states, the researchers must pay close attention to ethical considerations in the conduct of their work. Lau and Hayward (1997) used an action research approach in a study of their own to explore the structuration of Internet-based collaborative work groups in community health. The interpretations of the study suggest that role clarity, relationship building, information sharing, resource support, and experiential learning are important aspects in virtual group development.

3.3 Qualitative Tools

Qualitative methods allow designers to record explanations, perceptions and descriptions of experiences. Generally, a small number of individuals participate in a qualitative evaluation. Consequently, the results of this small number of participants cannot be generalized to the population. Some of the most common qualitative tools are; participant observation, diary studies, questionnaires, interviews and focus groups.

3.3.1 Participant Observation

Observing actions in ethnography is the main source of information. It may involve secondarily other field data collection tools such as informal discussions, individual and group interviews, documentary materials such as diaries, letters, subjects performed, photographs and video of actors, minutes, reports, newspaper articles, etc.

The ethnographic method can be decomposed conceptually into two research strategies: Non-participant observation and participant observation. In the first strategy, the researcher observes subjects at a distance without interfering with them; in the second strategy, a relationship is established in which the subjects and the researcher live and study together. In reality, this difference tends to disappear because we cannot study the social world without in some way being part of it. The ethnographic method therefore requires the researcher to participate with social actors, observing and simultaneously maintaining a sufficient distance to allow the researcher to maintain a balance between the two extreme attitudes that Davis (1973) defines as the 'March', i.e., those who exercise the greatest departure from objective analysis of the social situation versus the "converted" that is, those who tend to empathize completely with the people studied.

However participant observation provides a greater degree of active participation and therefore a medium to high involvement of the researcher. This position is described by Gobo (2001), in which the ethnographer is not content to observe and participate marginally in the daily activities of social actors, but tries to learn them and put them into practice. According to the same author the more appropriate attitude in the cognitive anthropological point of view is that of the 'outsider'. The ethnographer, ignoring many aspects of the culture of a group that he wants to access, tries to understand its conventions in order to act as a responsible member. Initially, the ethnographer takes little things for granted and has the ability to notice details that, in the eyes of the members, appear trivial and insignificant, or that are totally invisible. This attitude is much easier if he is culturally distant from the group. He is more like a classical anthropologist, who sometimes knew little of the language communities they studied.

In the field of ethnography, rigorous scientific techniques have been developed for collecting and analyzing data. An important contribution to systematic ethnographic observation in its early stages of information gathering, classifying and analyzing data is provided by sociologists Glaser and Strauss (1967) through their book on Grounded Theory, based on the idea of producing a theory inductively using only empirical observations.

During the observation, which should occur several times in different time slots to capture variable and repetitive behaviors, the observer needs to; detect the conventions, rules and regulations involved in various rituals; listen to the conversations of the participants; describe the characteristics of context, e.g., physical space and observable artifacts.

Schatzman and Strauss (1973) in particular recommended summarizing the notes taken during observations in three different sections: (i) Observational notes where actions and conversations are summarized in the most objective way as possible; (ii) Methodological notes containing comments and reminders of the researcher; (iii) Theoretical interpretations and reflections notes that they have developed. Corsair adds to these a further category of personal notes or emotional state of mind to record details of the researcher.

More detailed observation notes may be useful to refer to the three principles developed by Spradley (1980) for the drafting of ethnographic protocols: (i) Identification of the languages of the various actors involved; (ii) A verbatim transcript that truly captures the exact words used by the subjects in conversation; (iii) Description of actions as in a task-based analysis without recourse to abstract concepts.

Zelditch (1962) outlined three elements in an empirical approach to participant observation: The first element is the enumeration of frequencies of various categories of observed behavior, as in interaction analysis. Often there is an explicit schedule of observation geared to hypotheses framed in advance of participation. As Reiss (1971) observes, participation may lead to alteration of hypotheses and observation schedules; the attempt to observe systematically is ongoing. The second element in this approach is the informant interview to establish social rules and statuses. There may be systematic sampling of informants to be interviewed, content analysis of documents encountered, detailed illustrative incidents created and even recording of observations in structured question-and-answer format. The last element is an emphasis on systematic observation and recording of the milieu, where the phenomenological approach emphasizes the participant observer's experience of finding meanings through empathy. In the empirical approach, subjectivity has an important role, which is inherent to participant observation, with the attendant threat of researcher bias. That is, the researcher may be biased in what data are gathered and how data are assigned meaning. The participant observer may affect the phenomena being studied.

3.3.2 Diary Studies

Diaries are used for gathering user's experiences. There are various types of diaries in different formats such as written diaries, video diaries, mobile diaries, blog diaries, and photo diaries. Diaries are mostly considered as personal self-documentation tools. As Sanders et al. (2000) also indicated, there is no tool for diaries for group participation. Diary-like self-documentation, including digital technologies, enriches classical participatory methods and accounts for users' capacities, skills, and motivation. Self-reporting is an in situ method that takes place over time, allowing the exploration of new design contexts from the perspective of those whom future design may impact. Self-reporting methods take various forms; electronic sampling methods, diaries, and cultural probes (Larson and Csikszentmihalyi 1983; Carter and Mankoff 2005; Gaver et al. 1999).

Researchers have a handful of tools and techniques available for understanding everyday human behavior. But many of these techniques either require significant time and resource investment by researchers, such as contextual inquiries, or are divorced from empirical evidence, such as surveys. The diary study is a method of understanding participant behavior and intent that attempts to manage this gap by having participants' record events as they happen. This recording usually occurs in one of two ways: Participants answer predefined questions about events known as feedback studies or they capture media that are then used as prompts for discussion in interviews known as elicitation studies.

Field studies that require the researcher's persistent presence are difficult to scale. On the other hand, because of their reliance on participants to collect data, feedback studies have the potential to be scalable. However, participants are often reluctant to use them because the act of answering questions is a significant distraction from their main task. Also, because of the lack of an objective observer there is no way to verify to what extent logged information matches actual events. Media elicitation studies mitigate both of these concerns. In a media elicitation study, participants capture events, usually by taking a photo, and are asked about the event during an interview at a significantly later point in time. Thus for elicitation studies, capture is quick, and while the captured media still represents a subjective point-of-view, it has some empirical value.

Barsalou (1988) posited that episodic memory could be improved when a person is presented with cues about an event such as who was involved, where did it occur or what was done just before and after the event. However, while researchers have recently begun using diary studies using photo-elicitation, it is not evident how well media capture these cues and to what extent media facilitates participant reconstruction of events. Also, different media types will likely evoke different reconstructions and attitudes towards an event, but no study has yet shown how.

Carter and Mankoff (2005) have done research on improvements of diary study, thus contributing a tool to support media-based diary studies. They have found that images lead to more specific recall than any other medium, but that audio, in addition to making it easier for participants to capture information that does not have a visual representation, can be used clandestinely in situations in which participants do not feel comfortable using a photo to capture an event. As Graham and Rouncefield (2008) and Mattelmaki (2008) argued, the nature of participation in self-reporting has been the subject of discussion, with regards to probes in particular. Taking a participatory approach to self-reporting has largely meant two things: Meaning supporting active involvement and influence over design by participants.

As Hagen and Robertson (2010) state, doing a research in the context of social technologies allows perspective through which the 'personal stories' or 'data' produced through methods such as self-reporting, and the activities of creating them, can be interpreted, understood, 'read', or put to work. Such material is indicative of how the questions, topics or issues being investigated become relevant to potential future community members, as well as the forms and methods through which they may go about sharing those with others. That the material produced by self-reporting could potentially be put to more public uses raises a number of questions about privacy, consent and how data collection is framed. It also offers potential new ways in which participants can actively influence and participate in design through activities related to use early in the design process. Understandings of participation in self-reporting have largely focused on how much control participants have over how 'data' is produced and the degree of influence participants have over the interpretation of that material. As Hagen and Robertson (2010) state, managing these issues appropriately, using self-reporting studies as sources of seed content, could be an opportunity for future community members to directly contribute to the design of future platforms. Studies should be open-ended and participant-led, allowing participants control over "what and how" data is collected. In this way participants are recognized as experts of their own lives and are encouraged to choose "what and how" to represent their world. The second aspect according to Sanders (2006) is that participants play an active role in interpretation of the material that is collected as part of their ongoing participation in the design process as a whole.

The focus on self-reporting as a research activity is where individual participants record, reflect and share aspects of their lives with researchers. Different studies document self-reporting as a shared activity, such as Gaver et al. (1999) studies conducted with households, March and Fleuriot (2006) studies conducted with 'friendship groups', Isomursu et al. (2004) studies conducted with pairs. These collaborations include recruited participants as a formal part of the research design.

Hagen and Robertson (2010) propose that the spontaneous inclusion of others in the process of self-reporting reflects a sense of control and ownership by participants over the research process. Participants determine not just when and how documentation took place, but also with whom. For participants, accommodating the activities of self-reporting has always meant altering their daily practices to some extent. As Grinter and Eldridge (2003) state, the intervention of self-documentation facilitates reflection and at times behavior change.

In social technologies designed for community settings, contributors share stories, images and experiences around topics relevant to them. Hagen and Robertson (2010) found that in their 'Mobile Diary' method blurred the distinction between self-reporting and the production of user-generated content. This becomes possible due to the subject matter of the reports i.e., personal images, stories, and videos about a particular topic of interest, as well as the tools and format through which they were produced for communication, publishing and distribution. For example, Mobile Diary reports such events as a tour of a rooftop garden, home cooking experiments, or demonstrations of strategies for reducing household waste that told us something of participants' motivations and interests around sustainability, might also be ideal seed content for future-planned community sites around that same topic. Usually the content creation takes place after a system has been in some way formed and released to the public. The use of tools such as videos and camera phones early in the design research means the creation of seed content can begin earlier. As Twidale and Floyd (2008) state, opening up the potential for the structure of the future platform to emerge from the 'bottom-up'. For example, themes, navigation structures and taxonomies can emerge out of the content rather than be defined a priori.

Context mapping is a method that makes use of self-reporting. As Rijn and Stappers (2008) state, when looking at final research reports "users will automatically experience results with (their) personal expressions as their belongings." Their research looks at fostering a sense of authorship to the final reports that are created out of their research. Hagen and Robertson (2010) suggest that, when designing community platforms, there is also an opportunity for the material to be taken up in the design itself. Inviting participants to take the role of author and contributor, prior even to the development or specification of any particular platform, creates the potential for a greater personal connection between the design project and participant.

3.3.3 Storytelling

Storytelling draws on the approach used in narrative inquiry and supports exploration of the service idea. Through the use of simple words, the teller will illustrate the solution, as it is a story. This allows the communication of the idea inside the group and also the preparation of the first sketches for the storyboard. Storytelling is the conveying of events in words, images and sounds, often by improvisation or embellishment. Stories or narratives have been shared in every culture as a means of entertainment, education, and cultural preservation and in order to instill moral values. In designing for service, we need to understand people's expectations when they co-produce their service experiences. Storytelling is a method that can quickly reveal consistent patterns in people's experiences. Knowledge of these patterns can help designers produce ideas for services that have the best potential for resonating with their intended audiences.

A large part of our social life consists of sharing daily stories with other people. Storytelling is a key element of social interaction as it helps people to express their feelings and to establish a bond with others. Storytelling, in the sense of being able to relate recent and past experiences to relevant others, is considered to be crucial for the quality of life and psychological wellbeing of most people. Sharing stories is extremely difficult for people with limited verbal ability such as people suffering from Aphasia, or related chronic diseases that often lead to increased social isolation and depression.

The power of storytelling has been proved (Bate and Robert 2007) to be a relevant resource for design, to generate ideas and improvements and to challenge fundamental assumptions. A good driver for convergence is again the generation of a narrative, a vision or, in design terms, the building of a scenario (Carroll 2000). Scenarios and storytelling are often interconnected methodologies that have the powerful potential to facilitate convergence on distant futures and in complex projects if employed in a participatory approach (Jegou 2011).

3.4 Prototyping

As a collaborative and experiential method, prototyping has always been an important part of the Participatory Design toolset (Bødker and Grønbæk 1991). Examples in the following paragraph show that prototyping is becoming a living form of design research.

In the first example, Redhead and Brereton (2008) deployed an electronic-notice board prototype into a community. The prototype was then evolved in situ, in response to use and get community feedback. The authors reported a lack of success with traditional methods such as workshops, which were only attended by a few of the identified stakeholders (Redhead and Brereton 2008). Instead, installing a functioning prototype in a location that was physically shared by many members of the community (a local store), allowed people to experience the design as part of going about their daily lives. The authors saw this approach as a significant departure from earlier consultative 'Community Informatics' approaches—rather than seek consensus on intended use, stakeholders were able to indicate '*usefulness through use itself*' (Brereton and Buur 2008: 111).

Patchwork Prototyping (Twidale and Floyd 2008), an approach to the design of collaborative software, takes a similar approach, relying on the combination of open source tools, local code and mash ups of existing services. Rudimentary prototypes or 'patchworks' are pulled together and immediately integrated and used as part of daily practice: an easy way of supporting real user participation in actual use. Importantly, Jones et al. (2007) note that Patchwork Prototyping was observed as a phenomenon emerging out of practice, rather than being a method designed a priori. The researchers have since formed a research program around the approach. Botero and Saad-Sulonen (2008) also took a similar but deliberate "living research" approach in the development of the Urban Mediator software. In seeking to understand how social technologies could allow citizens a more active role in shaping council policies and responses to community issues, seed prototypes were used in a co-discovery process with the community. The Urban Mediator allowed citizens to track and contribute data about events in their city. Rather than undertaking traditional usability evaluations of isolated software components, Botero and Saad-Sulonen (2008) repurposed existing software to create 'concrete interventions' that could be co-evolved.

The approaches to prototyping 'in the wild' described here are possible because social technologies lend themselves to the deployment of simple prototypes that can be modified and evolved through feedback (Brereton and Buur 2008). Twidale and Floyd (2008) argue that such approaches only exist as a result of the current ecology of information technologies. The plethora of readily available and open source tools make rapid deployment and reconfigurations feasible and achievable. This supports Lievrouw's (2006) argument that reconfiguration is a key aspect of participatory design in the context of social technologies.

Floyd et al. (2007) described the advantages of such an approach in the following way, "The development proceeds and design decisions are made based on the users' collaborative experience of integrating the software into their every-day activities, not based on abstract design principles or predictions of what the users might need (p. 3)".

Through this experiential process both researchers and community members come to understand how such technologies become useful and meaningful in people's lives (Botero and Saad-Sulonen 2008). Participants are provided with a concrete and visceral experience of use (Twidale and Floyd 2008) as a way to evolve and participate in design.

As discussed, mobile technologies that enhance social interaction foreground a tight coupling between the practices of design and use because so much of their design takes place through use. In the approaches to design described here, this dynamic relationship is embraced as a design process in itself. Thus the practices of research and requirements gathering are combined with the practices of design and use, offering a way for members of the public to participate in design.

References

Barsalou, L. W. (1988). The content and organization of autobiographical memories. In U. N. E. Winograd (Ed.), *Remembering reconsidered: Ecological and traditional.*

Bate, P., & Robert, G. (2007). Bringing user experience to health improvement: The concepts, methods and practices of experience-based design.

Bødker, S., & Grønbæk, K. (1991). Cooperative prototyping users and designers in mutual activity. *International Journal of Man-Machine Studies, Special Issue on CSCW, 34*(3), 453–478.

Botero, A., & Saad-Sulonen, J. (2008). Co-designing for new city-citizen interaction possibilities: Weaving prototypes and interventions in the design and development of Urban Mediator. *Proceedings of PDC* (pp. 266–269).

Brereton, M., & Buur, J. (2008). New challenges for design participation in the era of ubiquitous computing. *CoDesign, 4*(2), 101–113.

Brydon-Miller, M., Greenwood, D., & Maguire, P. (2003). Why action research? *Action Research, 1*(1), 9–28.

Carroll, J. M. (2000). *Making use. Scenarios-based design of human-computer interactions.* Cambridge: The MIT Press.

Carter, S., & Mankoff, J. (2005). When participants do the capturing: The role of media in diary studies. *Proceedings of CHI* (pp. 899–908).

Davis, F. (1973). The Martian and the convert; ontological polarities in social research. *Urban Life, 3,* 333–343.

Dewey, J. (1986). *Logic: The theory of inquiry.* Carbondale, IL: Southern Illinois University Press.

Floyd, I. R., Jones, M. C., Rathi, D., & Twidale, M. B. (2007). Web Mash-ups and Patchwork Prototyping: User-driven technological innovation with Web 2.0 and Open Source Software. *Proceedings of the 40th Annual Hawaii International Conference on System Sciences.*

Gaver, B., Dunne, T., & Pacenti, E. (1999). Design: Cultural probes. *Interactions,* 21–29.

Gilmore, T., Krantz, J., Ramirez, R. (1986). Action based modes of inquiry and the host-researcher relationship, consultation 5.3 (Fall 1986): 161.

Glanville, R. (1982). Inside every white box there are two black boxes trying to get out. *Behavioral Science, 27,* 1–11.

Glaser, B., & Strauss, A. L. (1967). *The discovery of grounded theory: Strategies for qualitative research.* Chicago: Aldine.

Gobo, G. (2001). *Descrivere il mondo.* Carocci, Roma: Teoria e pratica del metodo etnografico.

Graham, C., Rouncefield, M. (2008). Probes and participation. *Proceedings of PDC* (pp. 194–197).

Grinter, R., & Eldridge, M. (2003). Wan2tlk? everyday text messaging. *Proceedings of CHI* (pp. 441–448).

Hagen, P., & Robertson, T. (2010). Seeding social technologies: Strategies for embedding design in use. *Design Research Society.*

Isomursu, M., Kuutti, K., & Väinämö, S. (2004). Experience clip: Method for user participation and evaluation of mobile concepts. *Proceedings of PDC* (pp. 83–92).

Jacobs, C.D., Haasbroek, J.B., & Theron, S.W. (1992) Effektiewe Navorsing. Navorsingshand leiding vir tersiêre opleidingsinrigtings. Geesteswetenskaplike komponent. Pretoria: Universiteit van Pretoria. Effective Research. Research Guide for tertiary training. Humanities component. University of Pretoria.

Jefferys, M., & Lashof, J. (1991). Preparation for public health practice: into the twenty-first century. In E. Fee & R. M. Acheson (Eds.), *A history of education in public health: Health that Mocks the Doctors' rules* (p. 325). Oxford, NY: Oxford University Press.

Jegou, F. (2011). Participatory scenario building at 'La Cité du Design'. In A. Meroni & D. Sangiorgi (Eds.), *Design for services.* Farnham: Gower publishing.

Jonas, W. (2005). Designing in the real world is complex anyway - so what? Systemic and evolutionary process models in design, In *European Conference on Complex Systems Satellite Workshop: Embracing Complexity in Design, Paris.*

Jones, M. C., Floyd, I. R., & Twidale, M. B. (2007). Patchwork prototyping with open source software. In K.St. Amant & B. Still (Eds.), *Handbook of research on open source software: Technological, economic, and social perspectives* (pp. 126–140). USA: Information Science Reference, University of Illinois at Urbana-Champaign.

Kemmis, S., & McTaggart, R. (Eds.). (1988). *The action research planner* (3rd ed.).

Larson, R., & Csikszentmihalyi, M. (1983). The experience sampling method. In H. Reis (Ed.), *Naturalistic approaches to studying social interaction: New directions for methodology of social and behavioral science.* San Francisco: Jossey-Bass.

Landman, W.A. (1988). *Navorsingsmetodologiese Grondbegrippe.* Pretoria: Serva. Research methodological Basic concepts.

Lau, F., & Hayward., R. (1997). Structuration of internet-based collaborative work groups through action research. http://search.ahfmr.ab.ca/tech_eval/gss.htm

Lievrouw, L. A. (2006). Oppositional and activist new media: Remediation, reconfiguration, participation.

March, W., & Fleuriot, C. (2006). Girls, technology and privacy: "is my mother listening?" *Proceedings of CHI.*

Mattelmäki, T. (2008). Probing for co-exploring. *CoDesign: International Journal of CoCreation in Design and the Arts*, *4*(1), 65–78.

O'Brien, R. (2001). An overview of the methodological approach of action research. In Roberto Richardson (Ed.), *Theory and practice of action research*. João Pessoa, Brazil: Universidade Federal da Paraíba (in Portuguese).

Reason, P., & Bradbury, H. (Eds.). (2001). *Handbook of action research: Participative inquiry and practice*. London: Sage Publications.

Redhead, F., & Brereton, M. (2008). Getting to the nub of neighborhood interaction.

Reiss, A. (1971). Systematic observation of natural phenomena. In Herbert Costner (Ed.), *Sociological methodology* (pp. 3–33). San Francisco: Jossey-Bass.

Rijn, H. V., & Stappers, P. J. (2008). Expressions of ownership: Motivating users in a co-design process. *Proceedings of PDC* (pp. 178–185).

Sanders, E. B.-N. (2006). Design research in 2006. *Design Research Quarterly*, *1*.

Sanders, E., & Stappers, P. J. (2008). CoDesign: International Journal of CoCreation in Design and the Arts, *4*(1), 5–18.

Schon, D. A. (1983). *The reflective practitioner. How professionals think in action*. New York: Basic Books.

Sanders, E., Brandt, E. & Binder, T. (2000). A framework for organizing the tools and techniques of participatory design, In *Proceedings of the 11th Biennial Participatory Design Conference*, Sydney, Australia, ACM.

Sanders, E. (2011). Sustainable innovation through participatory prototyping. *Formakademisk*.

Schatzman, L., & Strauss, A.L. (1973). *Field Research*. Englewood Cliffs, NJ: Prentice-Hall.

Spradley, J. (1980). *Participant Observation*. New York: Holt, Rinehart, and Winston.

Susman, G. I. (1983). Action research: A sociotechnical systems perspective. In G. Morgan (Ed., Vol. 102). London: Sage Publications.

Swann, C. (2002). Action research and the practice of design. *Design Issues, 18*(1), 49–61.

Twidale, M. B., & Floyd, I. R. (2008). Infrastructures from the bottom-up and the top-down: Can they meet in the middle?

Von Foerster, H. (1981). *Observing systems*. Seaside, CA: Intersystem.

Winter, R. (1996). Some principles and procedures for the conduct of action research, in new directions in action research. Ortrun Zuber-Skerritt (Ed., pp. 16–17). London: Falmer Press.

Zelditch, M. (1962). Some methodological problems of field studies. *American Journal of Sociology, 67*(5), 566–576.

Chapter 4
Project: Locast Health Diary

4.1 Application Area: Childhood Obesity

Obesity is a complex issue, with many interrelated factors involved in energy regulation and body weight. Ogden et al. (2006a, b) state that obesity is a significant public health problem and being obese depends on a variety of genetic, environmental and behavioral factors, and can lead to chronic diseases.

Causes, Incidence and Risk Factors of Childhood Obesity

Obesity is not only a medical problem but also has socio-psychological roots such as dietary causes, sedentary lifestyle, insufficient income, unavailable recreational areas or healthy food options in poor neighborhoods. Many authors (Ebbeling et al. 2002; Dietz 1998; Speiser et al. 2005; Kimm and Obarzanek 2002; Miller et al. 2004) state that a range of factors, which often act in combination, causes childhood obesity.

At the simplest level, one of the reasons for childhood obesity is an energy imbalance—children assuming more energy (calories) through foods and beverages than they expend through normal growth, physical activity and daily living. Excessive unhealthy snacking has been implicated in the increasing prevalence of obesity. Developing research suggests that the environments children live in have a profound impact on the foods they eat and the amount of activity they get. For example, most students have little or no time to be active at school, while unhealthy foods and sugary drinks are readily available.

Childhood obesity depends on various causes and risk factors that depend on individuals' social and psychological environment. The socio-psychological causes of obesity relate to dietary behavior, underestimation of food portions, sedentary lifestyle, family income, ethnical roots, accessibility of recreational areas and food services in the community, safety in the neighborhood, psychological factors, and genetic factors.

© The Author(s) 2016
P. Arslan, *Mobile Technologies as a Health Care Tool*,
PoliMI SpringerBriefs, DOI 10.1007/978-3-319-05918-1_4

(a) *Dietary causes*

Lack of access to healthy food options, available food services in the neighborhood, and the type of the chosen food might cause obesity. Eating inside home and outside home changes individuals' food choices in a significant way. It is not easy to have the occasion to eat at home due to individuals' tight work schedule, family members lack of knowing proper cooking skills and economic costs of buying fresh and healthy food. According to the study of Morland et al. (2002), in many neighborhoods—especially in low-income communities and communities of color—there is a lack of access to supermarkets, farmers' markets or other sources of affordable, nutritious foods. Yet, research indicates that when people have access to healthy food options, they consume more fruits and vegetables.

People consume so much food outside home due to its easy accessibility, availability, time saving and the pleasure of socializing. According to Zoumas-Morse et al. (2001), children today consume a significant amount of their daily calories away from home—either in school, at neighborhood stores or at fast-food restaurants. Restaurant meals can add twice as many calories and three times more fat than home-prepared meals. The study of Gordon et al. (2007) shows that many schools offer "competitive foods", those sold outside of the federally reimbursed school breakfast, lunch and after-school snack programs, which are often low in nutritional value and high in calories, fat and sodium. The most popular competitive food choices include cookies, candy, sweetened juice drinks and carbonated soft drinks.

(b) *Underestimation of food portions*

When you cannot measure food portions, it is almost impossible to estimate and compare them with other things. Especially obese teens or teens who are at risk of obesity, tend to underestimate their food portions. In the development phase of the Locast Health Diary project, nutritionists stated that underestimation of food is the most common issue in patients' weekly visits. Research study of Young and Nestle (1998) shows that people of all ages, education levels and body weights erroneously underestimate their food portions by 20–200 %. The United States Department of Nutrition provides an estimation of proportion by hand to ease people's food measurement habits in everyday contexts, to become familiar with the "estimating aids" and to practice their skills at home. These proportions guide their visual estimations of food measurements with a deck of playing cards, a computer mouse, a tennis ball, half a baseball ball, a Ping-Pong ball, cotton balls, an audiocassette tape, a light bulb, a 9-volt battery, and a standard-size hand.

(c) *Sedentary lifestyle*

Lack of physical activity is a contributing factor in many illnesses and diseases, especially obesity. Today, many children do not have safe places to play in the communities where they live, and few schools provide quality physical education or other forms of physical activity on a daily basis. Reduced physical activity also has been linked to a lack of recreational programming, poor air quality and safety concerns. According to the study of Troiano et al. (2008) more than half of children and adolescents are not getting the recommended minimum of 60 min or more of physical activity each day. Sedentary activities, such as watching TV, surfing the Internet or playing traditional video games are taking time away from physical activity. Delva et al. (2007) claim that the more time adolescents spend watching television, the more likely they will become overweight or obese. Children also are losing opportunities to be physically active during school hours. Due to shrinking educational budgets and competing academic pressures, many schools have cut recess and physical education. Fewer than four percent of elementary schools provide daily physical education.

(d) *Accessibility to recreational public areas*

Recreational public areas and services in the community are other important factors, which causes poor physical activity. Singh et al. (2009) state that children living in neighborhoods with the most unfavorable social conditions are 50 % more likely to be physically inactive; 52 % more likely to watch TV more than 2 h a day; and 65 % more likely to engage in recreational computer use of more than 2 h a day, as compared to children living in most favorable social conditions. Based on data (2007) children living in neighborhoods with no access to sidewalks or walking paths, parks or playgrounds, and recreation or community centers, have 32, 26 and 20 % higher adjusted odds of obesity than children in neighborhoods with access to these amenities, respectively.

(e) *Safety in the neighborhood*

Feeling safe in the neighborhood is another important issue, which affects encouragement of activities. Singh et al. (2009) states that children living in unsafe neighborhoods have 61 % higher odds of being obese and 43 % higher odds of being overweight than children living in safe neighborhoods (after adjusting for age and sex).

(f) *Psychological factors*

Emotional Eating is one of the biggest problems of childhood obesity. People create emotional attachment to certain types of food due to their taste, color, and packaging. As Karremansa et al. (2006) stated, thirst is a psychological goal. Being hungry and eating not only depends on physical hunger but also psychological and emotional situation of the person. Social isolation is another problem factor that leads to psychological problems and sedentary lifestyle. Himes and Thompson (2007) state that the glorification of the thin ideal and denigration of its opposite, an overweight or obese status, have been labeled "fat stigmatization". As Crandall et al. (2001) state, fat stigmatization stems from a variety of factors, including negative attitudes and cultural beliefs that equate body fat with gluttony and laziness, and the belief that weight can be controlled with self-regulation. According to Neumark-Sztainer and Haines (2004), Crandall and Moriarty (1995), substantial fat stigmatization research has accrued indicating that overweight and obese children, adolescents, and adults are often negatively stereotyped, treated differently, and face discrimination. Thompson et al. (2005) state that high body weight often leads to negative weight related commentary and teasing, and these experiences are strongly related to body dissatisfaction. In addition, according to Maranto and Stenoien (2000), Fikkan and Rothblum (2005), overweight and obese individuals receive less pay, are less likely to be hired in the workplace; Sechrist and Stangor (2005), and Gortmaker et al. (1993) state that these individuals' experience elevated rates of romantic rejection, and are less likely to be married.

Health Consequences of Childhood Obesity

The consequences of childhood obesity can increase risk for a wide array of health and economic problems. According to Ebbeling et al. (2002), and Lobstein et al. (2004), the increases in overweight and obesity among children internationally over the past three decades indicate that childhood obesity is a global "epidemic". Kopelman (2005) states that due to the rising prevalence of obesity in children and its many adverse health effects it is being recognized as a serious public health concern. Seagle et al. (2009), Rochelle et al. (2011), Wang et al. (2011) state that these children are developing "adult" diseases, such as type 2 diabetes and hypertension, and are at increased risk for heart disease, stroke, certain types of cancer and other serious chronic conditions as obstructive sleep apnea, depression, hypoxemia. Flaherman and Rutherford (2006) affirm that more than 100,000 children ages 5–14 suffer from asthma each year because of overweight and obesity. Being overweight in childhood increases the risk among men of death from any cause and death from cardiovascular disease. Researchers analyzing the previous data predicted future complications that obesity may cause. Bibbins-Domingo et al. (2007) state that researchers predict that if current adolescent obesity rates continue, by 2035 there will be more than 100,000 additional cases of coronary heart disease attributable to obesity. Allison et al. (1999) used data from six cohort studies in the United States to determine that obesity causes approximately 300,000 deaths per year.

Obesity has a substantial negative effect on longevity, reducing the length of life of people who are severely obese by an estimated 5–20 years. According to Koplan et al. (2005), the rising prevalence of obesity that occurred in the past 30 years is expected to lead to an elevated risk of a range of fatal and nonfatal conditions for these cohorts as they age. The study of Fontaine et al. (2003) shows that if the prevalence of obesity continues to rise, especially at younger ages, the negative effect on health and longevity in the coming decades could be much worse.

The medical expenses and indirect costs associated with obesity place a significant burden on our healthcare system. According to the Robert Wood Johnson Foundation Center Report (2010), obese children cost the healthcare system roughly three times more than the average child. According to the study of Trasande et al. (2009), between 1999 and 2005 there was a near doubling in hospitalizations of children with a diagnosis of obesity and an increase in costs from $125.9 million to $237.6 million between 2001 and 2005. Most of the insurance companies do not define obesity, as an illness so does not cover the cost of the camp. Insurance coverage, may be available if a physician refers your child to a camp to treat the host of obesity related medical conditions such as heart disease, diabetes, high blood pressure, high cholesterol, sleep disorder. Although in some literature obesity is not currently considered as "chronic disease". Medicare recently moved obesity out of a "not an illness" category. This is an indication that obesity will soon qualify coverage for insurance companies in the healthcare system.

Prevention of Childhood Obesity

To improve health from individual to policy level, it is important to provide the right environment and incentives to people to be able to make healthier choices, and provide them skills and knowledge to manage their healthy life. To prevent and solve obesity problem, there are strategies in action at five levels: (i) policy, (ii) environmental, (iii) community, (iv) schools, and (v) individuals.

At individual level, it is important that individuals make healthier changes in their habits, learning about appropriate portion sizes, consuming healthy snacks, learning new ways to manage stress or depression. Keeping a diary is one of the methods of tracking these activities consistently. A new study from Kaiser Permanente's Center for Health Research (KPCHR) (2010), funded by the National Heart, Lung and Blood Institute at the National Institute of Health, where the study is the one of the largest and longest weight-loss maintenance trials ever conducted, finds that one of the most important things you can do is write down what you eat. According to KPCHR research lab those who kept daily food records lost twice as much weight as those who kept no records. It seems that the simple act of writing down what people eat encourages them to consume fewer calories. Keeping a diary or sketching also helps people to express their emotions, fears, and dreams which is a way to cope with daily stresses of life.

At community level, it covers early learning and behavior for healthy diets, teaching home economics skills in shopping and cooking, and making available convenient healthy food to prevent "food deserts" in the suburban areas, where only relatively expensive and processed foods can be purchased. Achieving progress in these areas will require changes in communities. Not only does it change the availability and distribution of food at homes and schools, but also marketing and advertising practices. But it is also important, at a policy level, that the government provide adequate incomes to families adopting farming and food policies that encourage available local produce, put limitations on fast food and addictive substances, and require proper food labeling.

As Economos agrees "[We] started with a community that represents many urban settings in this country. Where there are children who are gaining weight unnecessarily because it's difficult to exercise, there aren't a lot of healthy food options, and quite frankly they have difficult lives. I think people are starting to understand that to affect health it's not just individual education and behavior change, it's creating healthy environments. And so we engage the police force, the city and the community members to really reshape the environment so that people can live a healthy lifestyle. With a $1.5 million grant from the Centers for Disease Control and Prevention, the city went about making changes big and small. They repainted crosswalks so people could see them better and feel better about using them. They hired more crossing guards for the schools. They say their efforts led to a 5 % increase in kids walking to school. They also tried to change the eating habits at school cafeterias."

Some good initiatives demonstrated successful scenarios in several ways. For example, the national school fruit scheme, which gives every four, five and six years old child a piece of fruit a day in school (in Europe five portions a day), was a real success, both in terms of take-up by schools and children and in influencing families and friends to eat more fruit. Moreover, teachers reported the benefits of this in schools in terms of both educational focus and children's greater ability to learn behave well and concentrate better.

To reverse the childhood obesity epidemic, it is necessary to balance the number of calories that teens are consuming and burning each day. This means making changes to the environments in which children live, learn and play that support healthy eating and physical activity. One part of the solution to reverse the obesity epidemic is making healthy, nutritious foods and beverages more affordable and accessible; discouraging the consumption of unhealthy foods and beverages; and achieving an appropriate caloric intake. Fighting obesity is not an easy issue. Current methods to prevent obesity vary from participative actions as attending support groups to medical solutions as decreasing the amount of intake nutrition by laparoscopic gastric banding.

Some of the common strategies are attending a fat camp, increasing physical fitness in schools, keeping a paper-diary consulted by a nutritionist, parent-child treatments offered by community health programs, and government healthy living campaigns. Fat camps provide tools such as well-developed programs around nutrition education, encouraging physical activities with non-competitive team sports, and age appropriate group discussions about emotional issues facing today's youth. Public campaigns and TV shows are alternative ways to develop awareness on obesity issues. Reality shows such as "The biggest Loser" and "I am too fat for 15" relate to obesity problems and provide information to help them change for a healthier life.

Government promotes research funds for participative community wellness projects conducted among universities, community groups and organizations. One type of government actions is support campaigns to fight the obesity epidemic, which is a future threat to citizens' health and nations' economy. For example, Let's Move! Campaign, launched on February 2010, is a U.S. nationwide initiative to engage and mobilize society towards solving the problem of childhood obesity. The program aims to work with national associations that have committed to advance body mass index (BMI) screening and assessment and new policies to support healthcare programs.

Some other government actions to provide support for a healthy life is putting higher taxes on sugar and soft drinks. In Denmark, the government imposed a "fat tax" on fatty foods to convince their citizens to eat healthier. In U.S., the state of Arizona proposed a program to increase the tax on sweetened soda and beverages, which includes "people who are obese or chronically ill, and those who smoke, would need to work with a primary-care physician to develop a plan to help them lose weight and otherwise improve their health. Patients who don't meet specified goals would be required to pay $50 under the proposal."

Another type of government action is to implement new regulations to provide clear and actionable information access to individuals, parents and caregivers for healthier decisions. The Task Force on Childhood Obesity Report (2011), a governmental act from the U.S. White House, includes this type of implementation of new laws and authorities enacted in 2010. The U.S. Department of Agriculture (USDA) and the U.S. Department of Health and Human Services (HHS) released 2010 dietary guidelines for Americans, which form the basis of all federal nutrition policy. These guidelines provide authoritative advice on good dietary habits to promote health and prevent chronic disease. For the first time, the recent dietary guidelines included a direct focus on obesity prevention and call for changes in food products in the marketplace to help simplify and translate this nutritional guidance for everyday use, which will be the "next generation food pyramid". At caretaker level, it is important to foster environments to support and access healthy choices for parents seek to make healthy choices for their children. The study done by Boutelle et al. (2011) supports this argument with a randomized controlled trial conducted between parent-only treatment and parent-child treatment. 80 families involved in the study with overweight or obese children aged to 8–12 years old, half of the families entered 5-month education program with kids in tow while another 40 families attended parent-only classes. At the end of a five-month term, researchers had measured both parents' and children's' body size as well as kids' daily caloric intake and physical activity. The study suggested that parent-only treatment could provide similar results to parent-child treatment in child weight loss and other relevant outcomes, and potentially could be more cost-effective and easier to disseminate. This shows how parents and caretakers are important in the weight loss of the child. Weight loss education for families is effective at helping children lose weight. Another study published in Health Psychology found that about one third of overweight and obese kids who attended educational classes in childhood decreased the amount that they were overweight by 20 % 10 years later, and 30 % were no longer obese. But this study indicates that when kids are directly educated without the involvement of their caretakers, they can still benefit from the given education.

At a policy level, government also provides initiatives in healthy eating and increasing fitness health promotion programs in schools. One example is enactment of the "Healthy, Hunger-free Kids Act" in order to improve children's access to healthy food in schools. The U.S. Department of Agriculture (USDS) took a critical step in releasing a proposed rule to update nutrition standards for meals served through the "National School Lunch" and "School Breakfast" programs based on the Institute of Medicine. The program aimed to increase fruit and vegetable at lunch and breakfast, increase whole grains served, and provide only low-fat or non-fat milk. Other governmental acts around nutrition information include a "menu labeling" provision requiring restaurants and vending machine operators. This action provides visible information about the calories in the food and front of pack labeling that people notice, understand, and use to make healthier food choices in their everyday life. Another USDA initiative "Healthier U.S. School" encourages schools to implement standards for the food served in schools, increase physical activity, and provide nutrition education. Physical activity is an essential component of obesity prevention, and a healthy lifestyle. However, only the integration of nutrition and physical activity programs provide a more comprehensive approach to childhood obesity. The Department of Education, Council on Fitness, Sports and Nutrition focuses on healthy eating as well as physical fitness limiting "competitive foods" sold in school and improving food placement and pricing in the cafeteria to support healthy choices.

For obesity prevention and the promotion of healthy living, previous academic studies and government initiatives above provide less technology-oriented but more participative and collaborative solutions. There is still a need to create new tools to tackle these complex problems using mobile technologies to create social interactions that people can live a healthy lifestyle.

4.2   Aim of the Project

Mobile technologies promote, enhance and enable advancement in managing childhood obesity and wellbeing of an individual. The Locast Health Diary project aims to provide a helpful set of tools for obese teens or teens at risk of obesity to record their socio-psychological environment and everyday health routines using these technologies.

The primary aim is to help teens to develop awareness of their daily habits and as a consequence, encourage them to change their behaviors towards a healthier life, since the process of self-reporting allows people to self-reflect and share aspects of their daily life. More in detail, the purpose is to evaluate if mobile tools as the video diaries, the location-based utility and the social platform can be useful to self-reflect on everyday habits, behaviors and decisions. As discussed by Grinter and Eldridge (2001) this process can also trigger participants to question their choices and everyday behaviors. The secondary purpose is to provide an effective instrument to nutritionists and educational experts to help teens to improve their actions in healthier behaviors. Psychologists Csikszentmihalyi and Larson (1987) state in their study that they have recognized the need to monitor patients in situ versus asking a participant to recall an event a week or more after the fact. Also the study of asthma conducted by Rich et al. (2000) shows that medical information gathering might be augmented by video diaries created by patients to show clinicians the realities of managing chronic disease in the context of their lives.

Locast, developed at the MIT Mobile Experience Lab, is a location-based civic media tool to enhance individuals' awareness of their everyday experiences while they are performing these activities and social dynamics taking place around them. The Locast system aims to understand obese teens' perception versus reality providing them tools to reflect on their problems and creating a network community in which they could develop awareness by social interactions. Video diaries, created by a mobile application, are used as a personal data collection system that allows individuals to record their eating habits, physical exercises and social activities during the day. The diaries, all recorded media as videos and notes, are visualized and shared in real-time on a location-based platform using mobile GPS receiver and location sensing technologies. The exchange of information affects health decision-making and creates a long-term behavioral change towards a healthier lifestyle. This web and mobile-based platform supports social networks where teens can share experiences within the community.

Locast as a platform could address obesity problem through using three components: Video diaries, social network and location-based map. Video diaries are important for recording and reflecting individuals' habits, choices and behaviors at the exact time of the activity preformed. The previous studies of Csikszentmihalyi and Larson (1987) show that factors can and do affect the memory of participants recalling past experiences. Recall can be caused by (i) the current emotional state of the participant, (ii) the length of time the participant is asked to recall, (iii) the participant sitting in a foreign environment. The video diary, which is a visual storytelling tool, could be useful to provide more "direct" understanding of individuals' experiences. This would not only help them to reflect on what they do in an actual day but also give further information about their everyday life, for future collaborative decisions with a medical expert, a youth counselor or a health coach. There might be privacy issues where the individual does not want to show himself or record his activities while he is in a social environment, however this could depend on the personality of the individual.

The second component of Locast system is the social network where social interaction between groups and peer-support evolves. There is a correlative relation between social support and physical health of an individual. According to Clark (2011) social support is one of the most important factors in predicting the physical health and wellbeing of everyone, ranging from childhood through older adults. The absence of social support shows some disadvantage such as the deterioration of physical and mental health among the impacted individuals. The presence of social support significantly predicts the individual's ability to cope with stress. Knowing that others value them is an important psychological factor in helping them to forget the negative aspects of their lives, and thinking more positively about their environment. Social support not only helps improve an individual's well being, it affects the immune system as well. Thus, it is also a major factor in preventing negative symptoms such as depression and anxiety from developing. Social networks provide the platforms to exchange support, motivate individuals socially being active and a part of a community. Being a part of a community makes individuals feel stronger and socially motivated to cope with their problems and also learn from others. However in chronic diseases, individuals take very personal their healthcare, therefore privacy and trust are critical issues to consider in the design of the social platforms which enables sharing, learning, supporting among individuals and groups. The level of sharing, control of their personal information and trust to other individuals can vary based on the personality, psychological factors, ethical and cultural background of the individual.

The third component of Locast system is the location-based map using location-based technology that allows individuals to understand the co-relation between place, which is their environment, and the type of their activity.

Individual's environment correlated with the location of the performed activity is an important issue to be considered in patient's chronic disease prevention. The location of the activity visualized on a location-based map help to sort out the type of activity and service located in the neighborhood. The change of individuals' location over a time period depending on their activity gives assumptions about the individuals' being active or passive during the day. All location-based information collected from video diaries aims to give additional information about the type of the activity and the choices that individual make.

Migraine, cancer, diabetes type 2, eating orders such as anorexia and obesity are all different types of chronic diseases. However, there are several reasons to focus on obesity as the application area of this project. The first reason is, obesity among other chronic diseases relates at a higher level with a socio-psychological environment of a patient and is preventable through healthier habits. On the other hand, diabetes has additional medical issues where patients need to use other medical devices to measure their blood glucose level, such as watching and evaluating their food. The second reason is, obesity is a social concern, making it likely that analyses will focus on environmental problems and community issues rather than as individual cases. To the contrary of eating disorders linked primarily with psychological problems, because their behaviors mostly derive from wrong information about food, obese people might be encouraged to change their eating behavior if they become aware that people are noticing and supportive. This brings up more problems of education and social issues, which make the Locast Health Diary study a good and visible change. The third reason is, Health Diary tools and Locast technology (mobile diaries, social network, and location-based map) could interfere with looking for opportunities to solve the obesity problem.

4.3   Methods and Tools

Locast Health Diary project is conducted through an action research strategy, as defined by Reason and Bradbury (2001), using a participatory narrative technique as mobile diaries and a location-based platform to record user environment and everyday health routines.

The main tools and methods used in the research, analysis and implementation of this study are: Video Diary, Social Network, Location-based Map, Visual Map, Cognitive Map, Video Analysis Map, Weak-Strong Matrix, Participatory Workshop, Pre-Post Project Questionnaire, Focus Group, Interview, Relations Map and Expert Analysis Card.

Pre-Post Project Questionnaire, and Interviews collected information from teens and experts in the research phase of Locast Health Diary. Taking the advantage of Locast mobile and web technology, video diaries, social network and location-based Map enabled active involvement of teens and collection of their behavioral, social and location data in the study. Visual and Cognitive Maps organized in three different states of mind which allowed us to make comparisons and support communication between participants during the focus group sessions. Video Analysis Cards helped to summarize and structure information for nutritionists.

The Relations Map, the network of core and extended experts and stakeholders, used in different phases of the project to interview subject matters, collect case studies, review scenarios, find participants for the study, and evaluate study results.

The Expert Analysis Card (Fig. 4.1) provided a template to collect interview results gathered from experts during the research phase of the study. The template captures two different type of information: The first column collects contact details and expertise area of the healthcare practitioner, whereas the second and the third column collect their feedback on the use of Locast mobile and web tools in obesity prevention, and suggestions on how to find participants for the implementation of Locast Health Diary project.

To collect data of everyday activities Locast technology, developed by the MIT Mobile Experience Lab, served as the main platform for both web and mobile application of the Locast Health Diary project. Locast mobile and web tools (Fig. 4.2) helped participants to collect their everyday data. Video diaries structured the recorded information in four segment-guided formats. Each segment asked participants a question or gave a task to perform. This user-generated content allowed them to sort out the type of activity, the environment they were in, and the mood. Social Network allowed participants to map out social interaction, and to share experiences with each other. The platform provided data on each participant's comment on friends' diaries, information about uploaded casts, and their perspective on "being healthy". Location-based Map helped to collect the location data of

Expert name **Susanna Barry**	How do find patients?	Social network
Title Program manager of Community Wellness at MIT Medical / Health Coach **31 march 2011 meeting**	• **Weightwatchers group@MIT** Donna Lagrotteria • **Shape-up Somerville** Jaime Corliss, Director of Shape-up Somerville • **Cristina Economos,** Associate prof. Friedman School of Nutrition Science and Policy, Tufts University	The kids aged between 16-17 almost want to show their own TV show, they want them to be watched by other people." They could be aware of their behavior looking at other people's behavior. Why should they be interested in participating in the social network?
Contact details Community Wellness, MIT Medical, E23-205 77 Massachusetts Avenue, Cambridge, MA 02139-4307 Tel. 617/253-3646 bars@med.mit.edu mit.edu/wellness	**Location based** "There might be a privacy issues. All kids that come to Pediatric unit of MIT comes from other places around Cambridge" WHY? They are not located all the time. Only the videos are located and the user decides it	**Video diaries** "...It lets know you more about everyday environment, feelings and emotions of an obese person. Because a healthcare provider will stay with the patient 20 min. in a room knowing only his physical problems but he has no idea the other 23 hours of the day are they doing?"
Expertise area Zan develops programs – for groups and individuals – on stress management, healthy sleep, and meditation for health, and eating and body image concerns. Her training is in education and developmental psychology, integrative health, wellness coaching, and therapeutic yoga. She was an author and co-author of articles on youth development, education, and health	Geographic link- "This can be an opportunity to map out after school teen centers, any high school teen programs in the community."	"Awareness of their own behavior" is the most underutilized health issue. Most of these people eat without paying attention to what they eat. They are addictive, autopilot and automotive. The video could be useful to "just-in time interruption". Being aware of "am I hungry, why am I at this spot?" this creates a mindful eating awareness of what they eat which leads to a behavioral change. (www.tcme.org - The Center for Mindful Eating)

Fig. 4.1 Expert analysis card

1. video diaries   2. social network   3.location based map

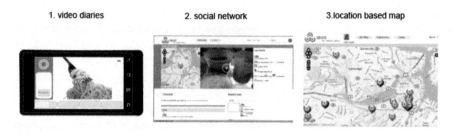

Fig. 4.2 Locast mobile and web tools

their daily activities, and the relational data between the type of activity and the location itself.

In the deployment phase, Participatory Workshops (Fig. 4.3) created an environment for discussion between participants.

Pre-Post Project Questionnaire (Fig. 4.3) structured with the continuous feedback from expert evaluation, and collected participants' feedback on Locast mobile and web application. In Pre-Project Questionnaire, the questions were dichotomous [Yes-No] to understand their background in using mobile technology, or experience in a media-related participative project. Open-ended, multiple choice and rating scales questions [4 scale—easy to difficult and 3 scale—agree to disagree] were used to capture the direction and intensity of attitudes to individual questions around technology, learning, user experience. In Post-Project Questionnaire, the questions structured under the themes of usability, learning, and user experience. Participants were asked to rate their degree of agreement with each statement on a 3-point scale with values from disagree (1) to agree (3). Statements addressed users' perceptions of system characteristics and experiences in interactive situations such as "I have learned a lot from watching the complete documentary", "I did not encounter errors or problems with the technology", "I was able to learn and fully understand the application", and "My perception of technology has changed after using the application".

Focus Group discussions (Fig. 4.3) structured around topics derived from the results of the video diary, social interaction and location-based data analysis. These group conversations with the involvement of experts in the room, helped to drive participants identify and conceptualize their future goals on healthy living.

Fig. 4.3 Participatory workshops, questionnaires, focus group

Cognitive Map, Visual Map and Pre-Post Project Questionnaire used to compare three different states of mind: Participant's perception, real state and participant's future desire. Both Visual and Cognitive Maps (Fig. 4.4) collected information on participant's perception, participant's real state and participant's future goal. Visual Map allowed teens to map out the activity type and their frequency of eating and physical activities. On the other hand, Cognitive Map allowed participants to map out all of their activities on a timeline, couraged them to think about co-relations of each activity, and the activity duration throughout the day. The Weak-Strong Matrix (Fig. 4.5) helped participants to analyze the difference between their typical day compared to their day using Locast tools, and identify their strength and weakness on each activity they performed during the deployment period.

Video Analysis Card (Table 4.1) collected information from the guided video templates of the personal diaries. This tool was mainly used by experts and researchers to extract information from the videos recorded in three categories: Environment, detail of activity, and mood. For each category, the card provided information about (i) reference image, (ii) observation of video content, (iii) participant answer, (iv) researcher's interpretation of video content.

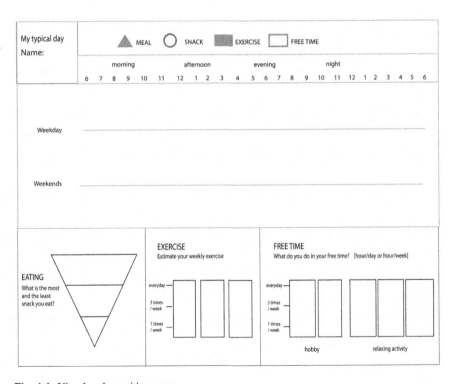

Fig. 4.4 Visual and cognitive maps

Fig. 4.5 Weak-strong matrix

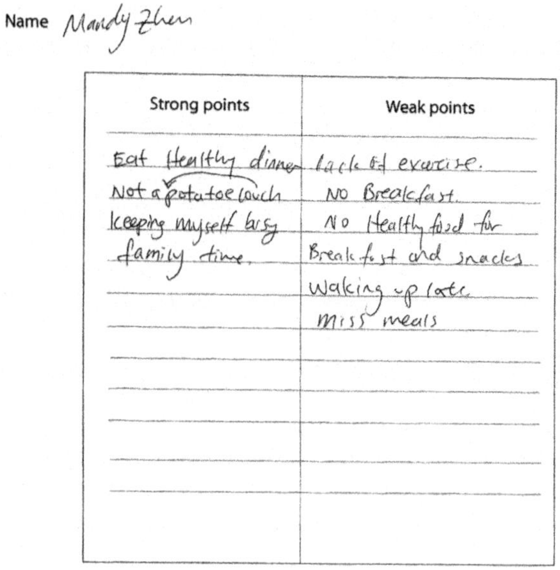

Name *Mandy Zhen*

Strong points	Weak points
Eat Healthy dinner	Lack of exercise.
Not a potatoe couch	No Breakfast.
keeping myself busy	No Healthy food for
family time.	Breakfast and snacks
	Waking up late.
	miss meals

Table 4.1 Video Analysis Card

Name of cast				
Activity type	Reference image	Observation	Participant answer	Researcher interpretation
Shotlist 1. Environment		Table 4.2		
Shotlist 2. Type of activity		Table 4.3		Table 4.4
Shotlist 3. Co-related activity		Table 4.5		Table 4.6
Shotlist 4. Mood		Table 4.7		Table 4.8

"Reference Image" is the representative image of the most important and critical recorded frame of the video segment; "Observation" is the objective information logged by the researcher and derived from the direct observation analysis of the environment, people and location of the video content; "Participant Answer" is the most important statement extracted from the participant's video diary; "Researcher Interpretation" is the researcher's subjective information directly observed from video-audio content, focusing on the participants' statement, body language, voice, and facial gestures. The Video Analysis Card allowed researchers to interpret observational data through guided questions to a structural analysis. The tool provided also flexibility comparing across activities of the same participant or different users with the same activity type.

Table 4.2 Shotlist 1. Environment—observation: Show the surroundings and introduce the place and people

Alone	Social environment
In a group	Not-social environment
Classmates	Home
Friends	School
Family	Gym
YES group members	Neighborhood
	YES

Shotlist 1. Environment. The first video shot aims to explore the physical and social environment of the participant.

Researcher purely observes the video shot, and fill in the template (Table 4.2) while looking for answers to these questions: (i) information about the name and type of location, (ii) if the participant is in a social or non-social environment, (iii) if he/she is alone or with a group of people.

Shotlist 2. Type of Activity. Second shot aims to explore the detail of the activity performed. Researcher purely observes: (i) the type of meal, and the brand if applicable, (ii) the accuracy of participant's estimation of the portion and ingredients of his meal, (iii) if participant estimates the ingredients or read nutrition label. Researcher also collects additional information about who is the cook of the meal, and whether the food contains high sugar or fat. Researcher uses different templates based on the activity type (Tables 4.3 and 4.4).

Shotlist 3. Co-related Activity. Researcher observes the existence of any co-related activity associated with the participant's core activity with the following questions: (i) if participant fully aware of his eating process, (ii) if this activity creates mindfulness eating or creates emotional eating (Tables 4.5 and 4.6).

For physical activity, the researcher observes whether participant produced something, or whether it is a relaxing activity or not.

Table 4.3 Shotlist 2a. Type of activity—observation: Explain the type of your meal. Estimate the portion and guess the ingredients. Check the nutrition label of the packaging

Type of meal			
Home cooked	Take-out	Packaged food	Restaurant meal
Left over	Fast food	Frozen food	School lunch
Name of food	Type of drink	Brand	
Participant tell ingredients		Participant do not tell ingredients	
Participant estimate portion		Participant do not estimate portion	

Table 4.4 Shotlist 2b. Activity detail—researcher interpretation: Explain the activity in detail. How long will you perform this activity? Are you performing in a group? Is it competitive?

Energy consume	Produce something	Creative	Expressive	Relaxing
Individual	Team	Individual in a group	Peer to peer	No

Table 4.5 Shotlist 3a. Co-related activity—observation: What are you doing while eating your meal or snack?

Participant does co-related activity	Participant do not do any co-related activity
Type of co-related activity	

Table 4.6 Shotlist 3b. Co-related activity—researcher interpretation

Over estimate	Exact measure	Under estimate	Do not measure	Does not tell complete
Participant guess ingredients complete [1–5 scale]				
1	2	3	4	5
Participants estimate portion right [1–5 scale]				
1	2	3	4	5
Socializing		Not socializing		Not clear
Emotional eating		Not emotional eating		Not clear
Mindful eating		Not Mindful eating		Not clear
Relaxing		Not relaxing		Not clear

Shotlist 4. Mood. Researcher purely observes the participants' mood (Tables 4.7 and 4.8).

In addition to four different shotlist templates, two additional free style formats are added to the guided video diary template: "Speak your mind" and "Community" (Tables 4.9 and 4.10). Researcher collects information to understand the reason why and in what situations participants would like to express themselves, and what communities do they relate to their experiences.

Table 4.7 Shotlist 4. Mood—observation: Are you hungry? How do you feel now?

Very hungry	Hungry	Not that hungry	Not hungry	Does not mention
Very satisfied	Satisfied			
Participant mood				

Table 4.8 Shotlist 4. Mood —researcher interpretation: How is your mood now?

Happy	Energetic	Relaxed	Excited
Sad	Tired	Stressed	Frustrated
Creative	Competitive	Productive	

Table 4.9 Speak your mind —researcher interpretation

Shotlist express yourself			
Express idea	Express emotion	After activity	Missing casts

Table 4.10 Community—	Name of community place
researcher interpretation	Type of activity that community place involves
	The reason why participant selects that particular community place

4.4   Project Development

The Locast Health Diary project lasted nine months and developed in a five-step process: Research, concept development, prototyping, deployment and evaluation. Experts and participants contributed certain stages of the study. Participants and researchers collected data and communicated with participants in each phase of the process using different tools and methods. New templates were created to reach different education levels and age groups, and to discuss different competences and areas of interests. Communication tools helped to reflect different perspectives of the project aim and provided a scale from simplicity to complexity of the content.

The research phase included primary and secondary research methods and lasted for three months. This phase included interview sessions with stakeholders and core expert team; collecting information from published papers and journals in the field of health psychology, social health, and obesity prevention; collected case studies from web sources, blogs, and vlogs; research projects collected from academic research groups laid out on the Relations Map; information collected from the attended conferences and seminars related to healthcare prevention and community wellness. The most time consuming part of the research was finding the participants and collaboration with actors for the deployment, which took six months. This time was mostly spent in learning the nature of finding actors and creating relationships with these stakeholders. The concept phase involved activities such as brainstorming with MIT Mobile Experience Lab team, scenario proposal and selection, scenario review and refinement with experts, and concept design for Locast mobile and web application. The prototyping phase included an iterative design and prototyping process for the Locast mobile and web application. The deployment phase included activities around several pilot tests and a 2-month application procedure to get permission for using a human as a subject in the study.

Finally the evaluation phase included the analysis of the study results, expert feedbacks, and follow-up sessions.

4.4.1   Research

Primary research involved discussions and semi-structured interviews with a core expert team whereas the secondary research involved data collection to understand user insights and needs. The Relations Map created as a result of all research activities performed in Locast Health Diary project. This network enables to reach out to different experts such as clinical psychologists, nutritionists, pediatricians, health coaches, researchers, design practitioners, health consultants, strategists, journalists, entrepreneurs, community leaders, and policy makers. The Relations Map aimed to fulfill various functions in the study: (i) content and concept reviews with multidisciplinary experts, (ii) find right partners to collaborate on the study, (iii) find right participants for the study, (iv) continuous feedback on the evaluation of the project outcomes. Furthermore, the map enabled grounded further collaborations.

Primary research
During the primary research phase, face-to-face meetings and semi-structured Skype sessions were organized to discuss the study with various experts. The objective of these meetings was to determine the application areas for a healthier lifestyle, prevention of chronic disease in a community setting, and ultimately how mobile diary tools (video diary, social network, location-based platform) could be mostly helpful to develop awareness for a particular chronic disease. Discussions and reviews from the experts were collected on the basis of (i) feasibility of the technology, (ii) tips and advice on how and where to find patients, (iii) usefulness for the treatment of the patient, (iv) usefulness for the physician to diagnose or write a treatment, (v) psychological benefit to the patient.

The Core Expert Team Interviews
Three medical experts participated as a core team to review certain stages of the Locast Health Diary study. The core expert team was composed of a clinical coordinator and a nurse practitioner at the Pediatric Department at MIT; a program manager and a health coach of the Community Wellness Group at MIT Medical Center; and a nutritionist at Children's Hospital in Harvard Medical School. They all have expertise on obesity, healthy living and community wellbeing, participated in scheduled meetings and semi-structured interviews to review conceptual framework. The core team was selected according to their available time and were all found to be easily reachable.

The core expert team provided feedback for content creation of the project and evaluation of the study outcomes. They discussed two main subjects: The first subject was the usefulness of Locast diary tools (video diaries, social network, and location-based map) for obesity prevention; and the second subject was how to find participants for the study. The interview sessions included semi-structured questions, and were registered with audio recordings. The researcher explained the functionality of the Locast technology, and showed an example of the use of this technology in other fields. The researcher later asked questions to trigger experts' imagination on how to apply Locast technology in the context of healthcare and wellbeing. An Expert Analysis Card with structured questions was used to elicit the experts' review of such topics as: (i) usefulness of video diaries, (ii) usefulness of social network, (iii) usefulness of location-based platform in order to solve in a healthcare-wellbeing context.

The Usefulness of Video Diaries. Experts stated that using video diaries could eliminate individuals' underestimation and unawareness in certain situations. The health coach contends, "Awareness of their own behavior is the most underutilized health issue. Most people eat without paying attention to what they eat. They are addictive, autopilot and automotive. Being aware of thoughts like 'Am I hungry, why am I here?' creates a mindful eating awareness of what they eat which has a good chance of leading to a behavioral change."

Some experts concluded that video diaries can reveal an individual's hidden realities. This insight could be useful for the expert in understanding that individual's thoughts can help them to develop their own awareness. One nutritionist stated, "When a doctor asks the patient a question, they might underestimate answers, but through video, the patient cannot lie and underestimate the fact." In addition the pediatrician added, "Through video diaries, it is less likely that they will lie or underestimate their portion. It is interesting to know what teens are thinking when they or their mom is making a meal, or what is their point of view and mindset. I would like to see their script. When I ask a question as a healthcare provider, they mostly underestimate or lie. However, video can't lie. The kid may say that he tidied and cleaned his room, but he had not. With use of a video, you would see if it is really tidied or not".

A nutritionist also added "Portion is important. They mostly don't get used to taking a photo of their plate; they forgot what and how much they eat. Most of the time patients underestimate the reality. During the doctor's visit, the traditional process is very time consuming. They fill out a paper, they think, they also may not give the right information but underestimate the fact." She furthermore discussed the role of social environment in healthcare with this statement, "Relationships in their social environment are very important for mental health or each activity they do. If you are being teased because you are overweight, or you choose that particular food because you are frustrated, you stop by a fast food restaurant on the way from school to home, because you are aroused by the smell."

She also mentioned the problems of traditional nutritionist models where they ask patients to log food diaries (food type and quantity) in a written format for three days. The first problem with this type of traditional approach is the underestimation of food portion by the person. The second problem, it is a time consuming process for the medical expert since she needs to analyze all the written information to make decisions for the person's health. Third, you cannot get any knowledge about their mood at the moment they are eating their food. The fourth problem is you do not have any information about their social family history such as how your parents behave toward you, or whether or not they force you to eat. During discussion sessions, the health coach stated that there were three words to help people set up their goal: mood, energy level and focus. She claims that it is very important to pay attention to what we eat while we are eating. She states, "Most people are eating mindlessly. While lunching, it is not good to be reading a paper, or in front of a computer, or on the phone, because we are not focusing on what we are eating and we do not feel with satisfaction the fullness of our stomach although we might eat more than we need."

Based on the expert statements derived from these interviews support that video diaries could be useful to gather further information about an individual's everyday life, where a healthcare provider could not know. One expert stated, "Video diaries let you know more about everyday environment, feelings and emotions of, for instance, an obese person. Because a healthcare provider will stay with a patient 20 min in a room knowing only his physical problems, but he has no idea of what the patient will do with the other 23 hours of the day."

All core team experts commented that in comparison to the traditional written diary method, video diaries should be less threatening and more fun due to its longer time commitment and required motivation throughout the day.

The Usefulness of Social Network: Our health coach commented that the age group between 14 and 18 years old is a good choice for an obesity prevention study. This age group likes to share and reveal their lives to other teens. She stated, "The kids between 16 and 17 almost want to show their own TV show, they want them to be watched by other people. They could be aware of their behavior looking at his or others video." The pediatrician stated, "Children share their thoughts and feelings. A self-motivating group, supporting each other might make them more active. They have ideas to be more active. It is curious for them what other people do, show them what they write, draw and meet."

She added, "Addictiveness of packaged food. Fast food, prepared packed cookies, is intentionally designed to eat one after another one. It is a layer of salt, fat and sugar that makes you hungry and go and buy another one, which makes you feel addictive. The packaging of these kinds of fast food designed attractive for you that make you feel hungry and make you buy."

Based on the expert feedbacks, a younger age group was selected for deployment of Locast Health Diary study. Since teens, especially who have problems with obesity, are still not aware of what they do regarding elders. They give underestimated information to their parents. Although obese teens might be shy and not open to share their daily habits and emotions, secondary research done on blogs, vblogs and social media websites shows that some obese teens are interested in expressing themselves, sharing their emotions and thoughts about the problems they confront in their lives.

It is a critical and highly important to think how to motivate teens to use these type of technologies and ultimately Locast system. Locast H2flow (Arslan et al. 2011) results showed that teens were likely to use technology and were more interested and eager to try new technologies. Creating a fun diary of everyday activities could interest teens as if they are the creator of their own narratives and stories.

The Usefulness of Location-based Platform. The core expert team stated that locations of individuals are important to understand their physical energy consumption throughout the day and the type of activity they associate with the location of the place. Health coach agreed that "This could be an opportunity to map out after school teen centers or any high school teen programs in the community, however still there might be privacy issues regarding location-based recordings." Nutritionist stated, "Location-based technology works in a city setting. In urban setting, mostly parents' drive obese kids to school or everywhere and they do not walk. Possible output from location-based map could be their calorie consumption, where are the best places they consume calorie, and where are the kids come from. They are all in same neighborhood or different places. This should be also considered when choosing the patient profile group."

Secondary Research on 'Video Diaries of Obese People'
What is the life of an obese person? How is living as someone extremely overweight? Social media reveals the feelings and problems that obese people encounter in their everyday life. Secondary research included the analysis of personal blogs, special interest groups, web-based diaries, online communities and analyzed to create a one-day story of an obese person.

From the analysis of these video blogs, diaries and forums, it was seen that young people are more creative and active and obese people, in particular obese teens, discuss and share multiple issues in these social platforms. Most bloggers write their diaries to keep track of their weight loss and share their emotions, and feelings within their anonymous network community. Motivation, self-esteem awareness, and family constraints were the common values that were found in these social contexts.

As a result of the secondary research on these social platforms, some obese teens found a necessity to share their feelings in an anonymous network environment. The reason mostly was to look for a motivational support, to express their feelings, and to get advice on certain medical topics. As an example from the blog, one obese teen was encouraged by another obese teen's yoga class.

4.4.2 Concept Development

After analyzing all research materials, initial concepts have been shared with MIT Mobile Experience Lab team members to discuss concept selection for deployment. From the analysis of research results and expert discussions, it is understood that there are three main important issues to prevent obesity: the environment, the type of activity and the mood of the participant. Video Intervention Assessment (VIA) method (Rich et al. 2000) that randomly collected data produces too much material to analyze since video data is not structured while recording the activity played an important role to design the Locast Health Diary tools. Video Intervention Assessment (VIA) method played an important role to design the Locast Health Diary tool. The tool randomly collects data produces too much material to analyze since the video data is not structured while recording the activity. Therefore, in the design of Locast Health Diary tool data is structured while individuals record their activities. Another input for the design of the tool was the prompt that forms the video creation template. For example, underestimation of portions was found as one of the most common problems in obesity, therefore in the activity detail of the video template, a prompt question about estimating portions has been asked, allowing participant to record that specific information.

Locast Health Diary Design Components

Mobile and web diary tools (video diaries, social network, and location-based platform) for obesity prevention are designed based on the existing platform of Locast technology developed by the MIT Mobile Experience Lab.

Locast technology is a framework that consists of four main components; (i) Locast Web: a user facing web interface, (ii) Locast Mobile: A mobile application developed by Android devices, (iii) Locast Core: The engine that powers Locast, (iv) Locast API: An API that lets clients perform all the core actions of the Locast platform. These components work together in completing the Locast experience. On Locast mobile, users are primarily engaged in content generation and information gathering. Locast web allows users to perform interactions. Locast Web provides browsing content and collaboration. The website has a map, tag and community browsing interface. Location-based Open Street Map [Google map] is used to present geographic content, with custom Javascript for performing Locast API requests load in content. Locast web requires initial sign-up and registration of the participants to be a part of Locast community. Locast Mobile's main goal is to capture the teens' day and share it with Locast Core, while categorizing the mobile captured data. For this reason, there are a number of entry points into Locast Mobile for the purpose of recording and sharing recorded media that is stored on the mobile phone. The mobile interface not only allows participants to capture read-only data but also permits them to edit their own media content. By allowing participants to delete, edit or review their captured information, it becomes a draft of the media content. Mobile application is structured to view "Activities" and "Videos". It has a feature that allows the participant to edit casts as drafts, store them until publication on the website when there is a Wi-Fi connection and view the casts that are generated and published. Geographic location is the core part of Locast, and on the mobile side, casts are recorded according to the captured location. Locast Mobile was designed to work regardless of an active network connection. It is used in situations where one is not guaranteed network connectivity, often due to factors such as expense and availability of connectivity. To accomplish that, it records all data directly to local databases and has a synchronization engine that can ensure that mobile centric data persists in the network when one is in range. Users can edit their previously synchronized casts (i.e. to refine a cast title or add tags to the cast) while they are offline with their Internet connection and those edits will be synchronized back as soon as the network connectivity is gained.

Locast Core is a platform that hosts Locast Web and Locast API, which are both web technologies. Django was chosen due to its powerful geographic data persistence, computation module and a powerful templating language for serving websites. Django integrated with video procession tools (i.e. mpeg) allows one to work with video content and to create thumbnails and transcoded copies of the videos. Locast API is a RESTful API, with a design goal to minimize barriers for developers. It uses HTTP verbs, such as GET, PUT, POST, and DELETE along with simple JSON data objects in order to manipulate data in Django. All actions such as GET, PUT, DELETE can be retried multiple times without a fear of destroying unrelated data. Casts are simply GeoJSON features, as are status updates and track logs. This allows easy integration with OpenLayers and OpenStreetMap in order to visualize our geographic content. For authentication, the API uses standard HTTP authentication to communicate credentials.

A special pairing system is developed to allow users to reduce the amount of information they may have to type on a mobile. The server generates a unique, temporary numeric code that is displayed on the user's personal Locast Web pairing page as both number and QR-code. They simply scan the barcode through a bar-code scanner or enter the numeric code into their phones, which the mobile client then sends to the server, and the server returns stronger, permanent credentials which the client can store and use for all future sessions. If users ever change devices, they can 'unpair' from the website and 're-pair' with their new device. Participants can personalize their content and be able to generate their "Cast Title", "Tag" and "The content of the video". Locast mobile application allows teens to capture and view their diaries; on the other hand web platform allows them to manage, publish, view and share their and other teen's diaries. Users can change their profile details as well as the context they provide in their diaries such as the tag names, location and description of their videos. Locast web platform allows teens to create a social community where it generates social interactions towards healthier discussion forums.

Locast Health Diary website is an important part of the system to develop awareness on teens where the platform allows viewing teens' their own video or other participants' video, and enables use of interaction tools such as commenting or liking others' everyday diaries. The website allows teens to view, like and comment on each other's video. The web platform includes a location-based map, where recorded videos are visualized based on the geo-location of their videos (Fig. 4.6). This allows mapping out the distribution of a single type of activity by different users, or all types of activities of the same user on the map.

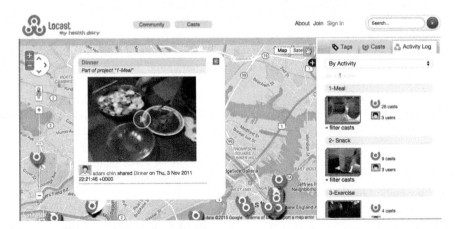

Fig. 4.6 Participant's video geo-located on a map

The video content on the web platform is organized in three levels: (i) data structure, (ii) tag management, (iii) filtering prioritize. Diaries are sorted out by most popular diaries, most recent diaries, and diaries in alphabetical order. Visual elements of interaction and semantics of the website are designed according to the defined concept.

To modify the desired interaction patterns at the front face of the web system, some of the non-functional parts of the website were hidden. Collaboratively working with a web developer, the web platform was designed to make the interaction easy to use, simpler to visualize and understand. The wireframes for the web platform indicate the placement of web elements such as where the logo should be, what elements should be in the navigation bar or on the side bar, the icons, indicators on the map and their interaction with each other.

There are five web page layouts on the web platform:

- Home Web page: This webpage includes all videos and activities recorded by participants.
- Community Web page: This webpage includes all profile of participants of the community.
- My Diary Web page: This webpage includes all "My Videos" of a participant. "Profile Info" and "Diary Timeline" streams all casts of a three-day diary of the participant.
- Single Cast Web page: This webpage includes single cast details and allows your interaction with the cast to edit, comment, like and view. If you click any cast on the webpage, it will browse you to this webpage.
- Activity Log Web page: This webpage includes the activity description and all the videos related to this activity. It also allows you to see the "Activity Playlist", a sequence of playlist of casts.

There is a semantic change on the Locast website to align with the content and simplify the understanding of teens. The recorded video is a mobile narrative with 4-guided template recording. The video includes four shots in which each shot is titled with given tasks. They are location-based narratives. The diary refers visual narratives, and all videos called 'a diary'. The Activity Log is the activity type performed by teens during their everyday life such as Meal, Snack, Exercise, and Chill Out. "The Speak your mind" and "The Community" are optional activities for participants to record additional information about their emotions, feelings and the environment they live.

The design of the web platform leans upon the social persuasion guidelines proposed by Tuomas and Oinas-Kukkonen (2011), which aims to increase the active engagement of users and a more user-friendly website. According to the social persuasion guidelines to support dialogues among participants, it is important to consider rewards, praise, reminder, suggestion, similarity, liking and social role in the study. Offering praises as prompts, a system can make users more open to persuasion. If a system reminds users of their target behavior, the users will more likely achieve their goals. Game rewards, altering users' media items such as sounds, background skin could be personalized based on user's performance. Some of these guidelines applied in the Locast Health Diary project.

For example, during the deployment phase the researcher posted comments, suggestions on the website which were seen on the activity timeline of the Locast web, and were also sent as an email to their mailbox. In order to support sociability, a person will be more motivated to perform a target behavior if he can use a system that observes others performing the behavior or, in social comparison, system users will have a greater motivation to perform the target behavior if they can compare their performance with the performance of others. In Locast Health Diary, participants were able to view others' diaries and post constructive comments each others' diaries. As social facilitation, system users are more likely to perform target behavior if they discern via the system that others are performing the behavior along with them. Activity timeline and given assignments on the Locast website helped participants create social facilitation. Recognition is another issue to create social support. In Locast Health Diary, the system allowed them to select the best video among all healthy eating, exercise and free-time activities, which results in setting challenges for competition among participants.

4.4.3   Project Locast Health Diary

In mobile interface, "Activity Log" and "Casts" are two main elements, which allow teens to create their own diaries and share them on a Health Diary web platform. Activity log includes categories to record everyday eating, exercise and free-time activities. Casts are recorded videos created by teens while performing any categories from activity log. On cast interface, all casts are categorized as published, unpublished and draft casts. Teens were able to edit, delete, view, re-record and rename all casts. Draft casts are videos, which are not published and are still in progress and could be re-recorded. Unpublished casts are videos, which could be published depending on the teens' decision. Published casts are videos that are located on the website to share with other teens. Casts are sorted regarding to descending time, from latest cast to earliest cast, to be able to modify the content easily at the latest. Instead on the web interface, on My diary page, the sorting starts from the earliest to the latest cast, to give the narrative format of diary from the first day to the last day of activity.

Activity log is designed to capture everyday activity type in participants' environment, which is mainly focused on eating, exercise and free-time activities. The content of a diary is composed of six primary activities (Meal, Snack, Exercise, Free Time), and two secondary activities (Speak your Mind and Community). Primary activities aim to collect information on daily activities, their mood and environment. Secondary activities support to collect additional information regarding to their everyday activities. Each activity on the mobile interface has been created through health diary mobile application (Fig. 4.7).

activity log   casts

Fig. 4.7 Interface of health diary mobile application

Primary and Secondary Activity List

1. MEAL. A meal is an instance of eating, specifically one that takes place at a specific time and includes specific, prepared food.
2. SNACK. A snack is (portable, quick, and satisfying) a small portion of food eaten between meals. The food might be snack food—items like potato chips or baby carrots—but could also simply be a smaller amount of any food item. (Excessive snacking has been implicated in the increasing prevalence of obesity.) It could be vegetables, nuts but also high-calorie, low-nutrient snacks such as cookies, ice cream, pie, crackers, chips, and chocolate. Snack as ice cream, pudding, cake, fruit, nuts, cookies, chips, candies, chocolate, fruit snacks, and granola bars etc. Drinks are as smoothies, vitamin water, ice tea, lemonade, chocolate milk, coffee and bubble tea etc.
3. EXERCISE. Exercise is any physical activity that enhances or maintains physical fitness, namely for one's muscle strength, cardiovascular system, and immune system as well as for enjoyment or to promote and maintain self-esteem. According to The National Heart, Lung, and Blood Institute report (2007) physical exercise is generally grouped into three categories: (a) Flexibility; such as stretching, yoga, Pilates etc. to improve the range of motion of muscles and joints (b) Aerobic exercises; such as cycling, swimming, walking, rowing, running, hiking, soccer, basketball, jogging, playing tennis and other activities that focus on increasing cardiovascular endurance, (c) Anaerobic exercises; such as weight training, functional, eccentric training (mostly in gym) and sprinting, all which increase short term muscle strength. In these types of exercises it is important to know if the physical activity is a team sport, individual or peer-to-peer activity, or if the activity encourages a competition/challenge, or not.

4. FREE TIME. Free time is time spent before or after compulsory activities such as eating, sleeping, going to school and mainly a time frame that is spent away from school. As Sharp et al. (2006) state that free time has potential for youth development, which is influenced by parental attitudes of interest and control, mediated by adolescent motivational styles. During free-time activities, teens can produce and learn new things, be creative, and relax. Free-time activities vary from relaxing to more productive activities. They include performing arts (singing, acting), creative pursuits (photography, photo editing, movie making, crafts, jewelry making, drawing, painting, cooking/baking, gardening), other hobbies (reading, browsing internet, watching TV, going to the movie theater, watching sports matches, meeting up with friends, playing cards, and video games), and after school classes and activities (extracurricular courses, dance practice, music lessons, art classes, and support groups).

5. SPEAK YOUR MIND. This is a self-expression template that allows teens to record their feelings, mood, and thoughts after activities like eating or exercise. Examples such as "I have eaten a lot. I feel totally full. I have eaten a huge box of rice chips while I was watching a film." or "I have eaten crunchy chocolate chips. I was satisfied with that for my afternoon snack." or "I ran for half an hour along the Charles River. This is a beautiful day. So many people are jogging. I feel very happy and energetic." or "I really love dancing. It just makes me happy and energetic." or "I could not help myself from buying all those delicious cookies at the supermarket."

6. COMMUNITY. Teens participate in an activity that explores their eating, exercise and free time activity experiences related to a place in the neighborhood. It is a free style template where teens explore their activity related to their community places in particular. Teens have selected two places of each in their community regarding to their most common eating, exercise and free-time activities.

The video content has been structured into four segments (Fig. 4.8), which includes information reviewed and required by nutritionist and health coach. The

Fig. 4.8 Guided video templates of health diary mobile application

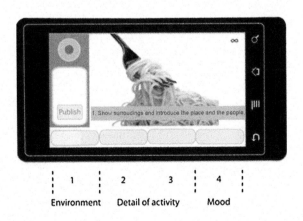

video recordings in the guided template set within unlimited time frame based on previous beta testing results, where participants could not complete their narratives due to limited time option.

Environment. Participants create a narrative about their physical and social atmosphere, describing the environmental factors, which contain information about the place and people in which participant is interacting or being surrounded with during the activity.

Detail of activity. Participants create a narrative of the detailed information their activity; estimate their food estimation portion and ingredients or the total amount of time spent on exercise or a free-time activity. The prompts appeared on the screen allow teens to be aware of whether they perform any co-related activity besides their primary activity, or whether they are eating emotional or automotive. Moreover, these prompts enables them to question if the activity requires any energy or focus on producing an artifact in a group session versus express themselves in a creative process alone.

Mood. Participants express their mood representing their emotional state (i.e. happy, sad, stressed, bored, energetic, tired, anxious, frustrated, lonely) at the moment while performing the activity.

Templates of shot lists are created according to the type of activities

'Meal and Snack' activities aim to explore teens' choices and habits on eating while they are performing the activity. These prompts in the meal/snack video template allows researcher to collect information around whether participants eat mindfully or emotionally, what their portion sizes are, and how they define a good and bad food (Table 4.11).

The first prompt aims to understand the environment the participant is in: Where do participants perform their activities? Is participant alone or in a social environment?

The second prompt asks: What type of activity and what estimates they can make of the portions of their food? What is the type of food? Is the food a home cooked meal, restaurant meal, take-out from a fast food chain? Where did the participants buy the food? Is the participant's estimate of the proportion and ingredients right? Is the participant underestimating the portion or ingredients of the food? If it is a snack, did participant read the nutrition label and reflect on it? Does the food contain high sugar or fat?

Table 4.11 Meal and snack shot list

Shot list for meal/snack
1. Show surroundings and introduce the place and people
2. Estimate the portion and guess the ingredients of your meal. Check the nutrition label
3. What are you doing while eating your meal?
4. How do you feel now?

Table 4.12 Exercise and free time shot list

Shot list for exercise/free time
1. Show surroundings and introduce the place and people
2. Explain the activity in detail
3. Do you produce something? Is it a relaxing activity? Are you performing in a group? Is it competitive?
4. How is your mood?

The third prompt aims to obtain more information about the activity: Is participant doing any co-related activity besides eating? Is participant fully aware of their eating process? Might this activity create an emotional eating or automotive eating?

Finally, the last prompt asks about their mood and emotion at that moment: What type of mood is the participant in? Does the participant's intonation confirm this mood? Is participant hungry or not?

'Exercise' and 'Free time' activities aim to understand how teens prefer to spend their free time activities. Video templates allow researcher to capture information around whether teens' are active or passive during their free time, or whether the activity they chose involves a team or competition (Table 4.12).

1. *Environment*: Where do participants perform their exercise/free time activity? Is the participant alone or in a social environment? With whom is the participant doing this activity? Is this a group activity with other people, friends, family, etc. or an individual activity?
2. *Detail of Activity: Activity Time Span:* How long will you perform this activity?
3. *Activity Outcome.* Did you produce something? Is it a relaxing activity? Are you going to produce something or express yourself? Will you be creative? Is the activity encouraging competition? Are you doing it in a group? Does the participant like the activity? Did the participant choose to do this activity or is it a result of influence or force by an outside source? Is it a group, peer-to-peer, or individual activity? Does it involve competition or a team challenge? Is it an energy consuming or relaxing activity? Is it a social or requires being alone activity?
4. *Mood*: How does the participant feel? Does the participant's voice confirm this mood?

'Speak your mind' and 'Community' are optional free style templates that provide participants the choice to express their mood or share their ideas about their activities. Researcher observes the video content of these two templates to understand why do participants need to express themselves, or which particular activities they do share a problem or a thought, ultimately whether their expression is a positive or negative one (Tables 4.13 and 4.14).

Table 4.13 Speak your mind shot list

Shot list for speak your mind
1. Express yourself

Table 4.14 Community shot	Shot list for community
list	
	1. Relate your experience within the community

Some of community-related prompts are: "Where is your favorite place to walk? What resources are available in your community for exercise or free-time activities? Where is your favorite place to have breakfast with friends? Where is your favorite place to go out to eat in your neighborhood? Do physical exercise places in your neighborhood encourage you to do physical exercise? Do you attend any classes or groups in your community to improve your health? How do you motivate yourself in your community to be active? Where in your local community do you go when you are hungry? Where do you go in your local community to do an activity to exercise or relax? How do people in your community try to be healthy?"

4.4.4   Deployment and Evaluation

Deployment and Evaluation phase is composed of two sections: First section explains the legal issues regarding the use of human subjects as an experimental study, and second section explains how to find participants for the deployment of study.

Legal-policy issues regarding deployment with human subjects
A Committee at MIT reviewed the Locast Health Diary study on the use of Humans as Experimental Subjects (COUHES). In order to conduct a research on human subjects, the researcher needs to know about ethical and legal guidelines to encompass all types of interactions with human subjects including direct contact, indirect involvement, and analysis of data.

Locast Health Diary study included documents such as (i) The basic information about the project: Funding, information about principal investigator, title of the study, human subject training, and dates of research, (ii) Study information: Purpose of study, study protocol, descriptions of tools and procedures to be used, (iii) Human subjects: Subject recruitment, compensation, potential risks, potential benefits, data collection, storage and confidentially, informed consent involved in the research project and investigators insurance, consent forms to participate in study, in interview, assent to participate in research for minors below 18 years old age, parental consent form, recruitment text to put ad outside, questionnaires.

The Process of Finding Participants
According to the Green et al. (1995) guidelines, various questions are considered before involving a stakeholder in Locast Health Diary project. These questions address the nature of participants' involvement, and the purpose of the research. Community participants were enabled to contribute their intellectual resources as their expertise to the research process. The deployment was an effort to research issues supported by members of the defined community. The purpose of the

research facilitated learning among community participants and resources external to the community and empowered the community to address social determinants of health.

From process, context and methodological considerations the research process applies the knowledge of community participants in the phases of implementation and evaluation. For community participants, the process allows learning about individual and community health issue. The process allows flexibility or change in research methods and focus. Community participants become involved in analytic issues: interpretation, synthesis and the verification of conclusions. The research process reflects the defined community for action. The process reflects a commitment by researchers, experts and community participants to social, individual or cultural actions consequent to the learning acquired through research. Community participants benefit from research outcomes. There is an attention to, or an explicit agreement for acknowledging and resolving in a fair and open way any differences between researchers and community participants in the interpretation of the results or respect to ownership of the research data.

Finding participants was one of the most difficult parts of the Locast Health Diary study. There are various ways to find subjects to participate in a human subject research. However, the difficulty arises when especially looking for obese teens. Since the target group relates to obesity, it is not only difficult because they are obese or at risk of obesity, which in most cases they feel embarrassed, anti-social and isolated from the society, but also obese teens for which there might be positive and negative sides of being in an adolescent age. Therefore, teens that are a part of an obesity prevention program have been involved in the experimental study. Since the request comes from the researcher side, it is important to find ways to convince participants to be a part of the project.

Two approaches were followed to find subjects for the study. First approach was to contact subjects through various research or governmental institutions, community groups, medical experts and university hospitals. Second approach was to contact subjects directly without any intermediate person or institute. Although the first way is a more secure and trustable way to find subjects, it is at the same time the most difficult where issues may arise, such as necessary agreements, loss of information, conflict of interest, and necessity to find common goals for a win-win situation. There were two possible directions to collaborate with institutes: One was collaborating with a medical expert and its patients; other one was to communicate with community group and teens. The Locast Health Diary study deployed with a community group due to time limitations and privacy and policy issues of deployment with other stakeholders. There were reasons to choose to deploy Locast Health Diary study in a community setting rather than in a hospital setting. First, deploying the study in a community setting helps to analyze the social influence among participants whereas in a hospital setting the interaction happens between patients and doctors remains medical oriented rather than socio-psychological level. Second, the collaboration with a hospital rises patient privacy issues, which requires both sides approval from Intellectual Review Board (IRB), and extends the deployment time. Last, community leaders are more innovative and their aim is education therefore more open to new activities, whereas in a hospital setting, the doctor is responsible for the patient and the priority is curing the patient.

First approach is contacting subjects indirectly through a mediator. The initial approach was contacting subjects through a mediator, which could be healthcare providers, healthcare researchers, community group leaders, director and coordinator of community groups, teacher and group leader of students. There are various advantages to select this approach: (i) it is important to involve more actors in the project to create a socially and economically sustainable system, which results in future collaborations. (ii) it is easier to replicate the proposed research in other contexts or with another group of people together with the collaborated entity. (iii) it is easier to do an action research where the research itself needs a cycle of various design process and easier to call people and redo the research multiple times —point of reference. (iv) creating a network of actors, which will be useful for future research references. However, there are also disadvantages: (i) contacting subjects indirectly through various actors slow down the process and it is also time and energy consuming. (ii) depending on the type of stakeholder the project direction might change due to different interest and goals of both entities. (iii) IRB agreements that have to be approved by both entities if the stakeholder is a part of a university or hospital. (iv) conflict of interest and expectations, (v) trying to find a common goal and interest for a win-win situation.

Second approach is finding subjects with the help of the core expert team. The core team provided suggestions on how to find participants, and followed a protocol for the study of Health Diary. They gave contact names from healthcare research centers, community initiatives, and government, which have expertise in childhood obesity research. The stakeholder map, created by Omni Graffle diagramming application, was useful to map out the possible stakeholders to contact for the study. In each circle of the stakeholder, the name, title and the institution was indicated. Primary or secondary arrows identified the relations. Dark colored arrows indicate the primary relation; dotted arrows indicate secondary bonds with the stakeholder in the network map. Contact names varied from a local gym where all kids go to exercise (YMCA), to a more specialized gym that helps obese kids through running a losing weight program (Bodies by brandy), or from a hospital research group to high schools in the neighborhood.

To recruit the right participants for the study, researcher reached out to different stakeholders with different presentation templates. In most cases, research groups in a hospital setting showed higher interest than community groups participating in the research project. One of the main reasons for this could be the motivation of scientific contribution by the research groups versus proposing a framework for the children at a community group by non-profit organizations or government institutions. Another result was to find subjects for the study from a school was more difficult than in a non-profit community group or a local gym.

Second approach is to contact participants directly through putting an ad in public spaces or posting ads on Internet. Therefore, to reach the right participant group, we put a text-based ad with a date, explaining the aim of the project and the contact information indicated with approval by Institutional Review Board (IRB).

The advertising was put at common places on the MIT campus, Pediatrics Department billboard and emergency room at MIT Medical Center, distributed to colleagues of a core expert team.

Since the project includes a concern of handing in mobile phones for a period of time, the privacy and security issue with the mobile phones, it is highly important to contact trustable and reachable participants. In particular, when participants are teens it is more logical to contact with their parents, teacher or their responsible contact person, who represents an authority that teens might feel responsible to. It was not easy to reach teens by advertising on billboards or putting an ad in common public spaces—it was less successful than getting in touch with a group at a hospital or in a local community. The reason why teens would not be interested in the ad could be: (i) the subject might not find any information on the ad regarding what the project looks like. (ii) the way to communicate with subjects in the advertising, (iii) not suitable place to look for obese patients. Another option was to recruit teens through online or offline support groups, local newspaper or reaching out through a social network of an obese person, however it still rises security and trust issue with the participants to whom the researcher would be handing mobile phones. One of the important part of recruiting the right participants for the study is good and clear communication of the project aim and understanding of their needs. Another way to recruit is to get a list of patients from the medical doctor and send them an email. However, since it is privacy information of patients, medical doctors in the hospital cannot provide this type of patients' information. The only thing they can do is to show their patients the ad that is put outside without any encouragement. Another obstacle is the tight schedule of clinics that makes it hard for medical experts to assist with advertising or informing patients about the research project.

4.4.5 Locast Health Diary Study

The Health Diary deployment lasted four weeks including three workshop sessions of one and a half hours. The deployment phase was an iterative process, where problems occurred, solutions found and decisions made in order to reach the final goal.

Locast Health Diary deployment had three objectives (Fig. 4.9). First objective was to explore if the workshop results evaluate to develop awareness and push teens to think towards a positive behavior change, which leads to prevention of obesity. In particular, the main aim was to develop awareness through self-reflection of viewing your own video and others' video and interacting each other on social platform, or location-based, or recording video diaries. This is being evaluated by discussion questions and the interviews with experts, video diaries, maps and questionnaire.

Second objective was if Locast system is capable of creating interactions to stimulate developing awareness; if video diaries, social network and location-based platform are useful for developing awareness towards obesity prevention. Eating,

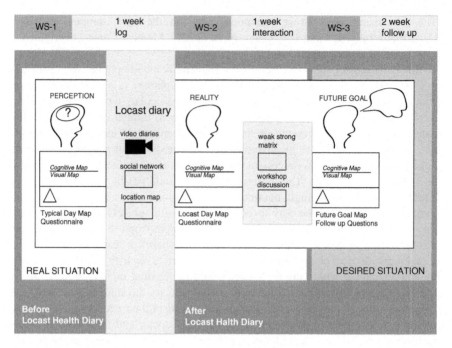

Fig. 4.9 Locast health diary study

exercise and free-time activities are mapped out to evaluate the distribution of the activities on the location map. Commenting, viewing and liking patterns on the website are analyzed to see whether those interactions help to create awareness and think about to change behavior towards healthy everyday choices.

Third objective was to create guidelines for other actors of healthcare system such as designers, medical experts and educators to apply these methods in their program. This is because if you design a concept by bringing all objectives and needs together, you need different frameworks to meet these goals, and providing a guideline would help them use designed tools to reach their goal, but at the same time give them direction towards a common purpose. For example, 'Video Card Analysis' could be useful for a nutritionist or a medical expert to analyze whether the participant is an emotional or mindfulness eater, which is one of the most causative factors of obesity, or map out activities in the neighborhood. This way, the stakeholders are empowered by these tools, and streamlines with a clear direction the way they analyze data and help participants.

Developing awareness is the relation between perception, reality and the future. To analyze the fact that if Locast Health Diary tools create awareness or not, it is important that participants understand the relation between their perception versus reality which result in identifying future goals. In Locast Health Diary project, video diaries, pre-post questionnaire, and cognitive-visual maps are used to understand the relation between participant's perception versus reality. To transfer from real state to a future state, focus discussion groups with the analysis of weak-strong map, and the cognitive-visual map help participants to reflect back to their three state of behavior: perception-reality-future location-based post project.

"Teens going Healthy" Program at YES

The Locast Health Diary study is conducted with participants from a "Teens going healthy" program at Youth Essential Center (YES), a community-based organization that aims to promote health of Asian teens in Boston. "Teens going healthy" is an obesity prevention program, supported by the Community Health Group at Tufts University, that aims to provide funds for innovative programs to address the health needs of the Asian community. The prevention and intervention program addresses specific issues faced by youths and their families and provides trust, self-confidence and self-esteem for teens. Six teens, two female and four male, participated in the Locast Health Diary project for a month together with an expert team of a nutritionist and a youth counselor. Teens were aged between 14 and 18 years old, which were a part of "Teens going Healthy—an obesity prevention program" in Youth Essential Center (YES), promoted by Boston Asian and the government. The youth counselor of the program selected six teens that might likely be confronted with obesity problems for the Locast Health Diary study. Among the selected ones, some of the teens go to the same high school, and some others live in the same neighborhood.

"Teens go Healthy" is promoted as well-being program to teens to receive their active participation, but it is an obesity prevention program. The program has 40 teens in total, where most of them are voluntarily participating and some of them are encouraged to participate by their parents, since it is a kind of after school activity for them. The weekly curriculum of this program covers several group and individual one-to-one discussions led by youth counselor, and tracking of their progress. Core teens come to YES once or twice a week and non-core teens come from time to time. The currently used methods in the program are regular meetings with core teens twice a week and appointment-based meetings with non-core teens. Educational health workshops and activities such as a field trip to a supermarket to show teens healthy options, or provide nutritional information of how to switch their cheeseburger menu with a healthier lunch option, are also important part of this educational program.

Locast Health Diary Study Workshops

Three workshops were organized during the study of Locast Health Diary. First workshop introduced the project, explained Locast system, and collected participants' profile information. Second workshop provided more detail on how to interact with Locast website and allowed participants to do a test uploading their videos on the website. Final workshop created a discussion environment for teens to self-reflect on their behaviors, compare their perception versus reality, and design their future goal towards a healthier lifestyle.

Intro Workshop. Introductory workshop aimed to provide information to participants on the project aim and how to use the Locast technology. Each participant received a mobile phone featured with the Locast application to do a trial run recording and publishing a video. All parental, participation and minor consent forms have been collected, and check in-out consent forms have been signed by each participant for the security of the mobile phones. Teens completed pre-project questionnaire and designed their typical day map, which helped the researcher to collect data about their perception of everyday eating, exercise and free-time activities. 'Typical day' map (Fig. 4.10) helped teens to map out the type, routine and frequency of their daily activities where pre-project questionnaires collected (i) participants' contact information (ii) description of participants' typical daily activity routine written in a narrative format (iii) their preferred community services to perform these activities (iv) diary experience on tracking their activities (v) technological background and familiarity with mobile applications, recording video, and social networks.

In the workshop, participants were divided into two groups. Tutorial materials helped teens to understand how Locast mobile and web application work. They were tutored to sign-in the Locast website and paired their mobile phones to test how to record, re-record, view, delete, edit and publish a cast on the mobile phone.

After the introductory workshop, teens started to record their daily eating, exercise and free-time activities over a week. During the weekday, they recorded their daily routine activities between school, home and after school places, whereas on the weekends they mainly record activities around and outside of their community.

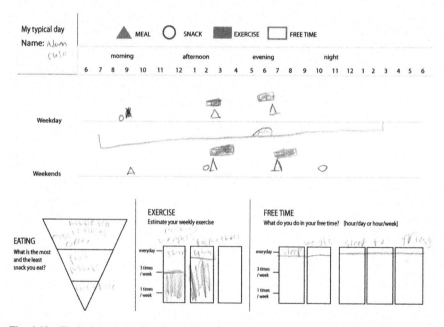

Fig. 4.10 'Typical day' visual and cognitive map

On the first day of logging, we had two problems. First one was a problem with the time-stamp of the videos published on the website. Time-stamp showed Greenwich time where the Locast system converted it and posted it on the website as Eastern Time. To solve this problem, we did several trials with two spare mobile phones, downloaded Locast application and published casts both from admin's account and test user account. Second problem was with the mobile phone's ability without a SIM card trying to get the location of the video on the website. During the study, the location tagging of each video did not show up in any cast, or it was placed with a wrong location tagging on the website.

To solve these two technical problems, the researcher posted several how to do videos on the homepage of the Locast website to reach all participants. These videos included informative videos, explaining step-by-step solutions to technical problems, self-motivating videos, and feedback videos of the recorded diaries. During the study, it was observed that participants did not pay much attention to the community template. As a solution the researcher posted a motivational video, explaining how to utilize the template.

Interaction Workshop. After completing their first week of activity logging, they checked out the given mobile phones. During the interaction workshop teens uploaded all their videos to the Locast website with the assistance of the Locast team, checking the accuracy of their collected videos' location and time-stamp. After this first check by the MEL team, teens went over all their video and they re-tagged the correct location and added the correct time-stamp for the ones, which were not correctly published.

Some teens had missing activities during this one week activity-logging week. One male participant indicated "He did not record any video, because he did not eat anything for two days. He just drank water." Although we can not prove what the real reason would be behind this fact only by analyzing video, whether 'he did not' want to say that he did not log his two days' or 'he does not want to tell the truth', but this workshop helped the group to create a discussion around this topic. This particular case identifies one of the differences between a written diary and video diary. In the written diary, teens might write the name of the food they have eaten, but the written text does not necessarily give us information about their environment and mood unless they describe it. The control of how much and how detail to provide information depends on the participant. However, with video diary there is both visual and auditory feedback, which might provide us more information about their behaviors. By using this method, participant owns the control how to shape the video content, although the guided video template helps the participant to structure the collected information as well as capture the emotion, environment, and mood.

Final Workshop. The final workshop aims to make teens self-reflect on their diaries, and discuss together with other teens how to create awareness of a healthier lifestyle. This participatory workshop created discussion to make them compare what they think they do and what they actually do in their everyday lives. As a result of this, they defined their problems, which helped them to construct their future goal towards healthier habits. Final workshop structured in three sessions.

The first discussion session focused on discussing the utility of Locast website. Teens compared each other's diaries that are shared on the website and evaluated the interactions that occurred within the web community. The second session focused on teens comparing diaries, and the final session helped teens to create their "Future Goal" map.

First session had a three-task assignment to encourage participations involvement in the discussion and self-reflect on their diaries. This assignment provided additional information that could not have been collected by teens during their one-week interaction period. First task was to look at each of them and tag their own videos with one of these words: 'healthy', 'not healthy', 'I don't know'. Second task was to select the best video that represents 'best healthy food experience' and 'best exercise experience', and explaining why they think it was the best. Third task was to let teens think their weak and strong points looking at all the diaries on the website, and make them think about what behaviors or choices to change in their life. This assignment allowed teens to focus on the content and type of their single activities throughout the day.

Second session was to make teens compare their "Typical day" map and "Locast" map; and let them discuss differences between their perceptions versus reality. "Typical day" map was created to capture their perception on teens' behaviors in the first workshop, whereas "Locast" map was provided by the researcher, created from the analysis of the collection of videos that are published on the Locast website. Teens compared their own videos looking at the time, the type and the quantity of activities recorded throughout the day. After looking at their own maps, they looked at other teens' maps to question what might be the problems in those activities.

The first two group discussions encouraged comparing teen's "Typical day" map and "Locast day" map. The researcher and nutritionist led the conversation through discussion questions: "When you compare your maps, what differences do you see?" "Do you eat better in the morning or afternoon, viewing from the videos?", "Where do you eat the healthiest do you think, watching your own video?", "Were you happy with your eating, exercise and free-time activity choices throughout the week or weekend?", "Which particular activities and moments were you not so happy about or would like to change?", "Were all your activities related to your location near by school or home?"

These questions helped teens to reflect on their problems, in the choices they made or behaviors they had during the Locast week.

In the last session of the final workshop, teens defined their problems, and evaluated their strengths and weaknesses, and designed their "Future Goal day" map, modifying their weak points into stronger ones with specific, measurable, and realistic goals. To drive the conversation, researcher asked questions such as "Considering your weakness and strength, what you would like to change in your life? How could you improve your day for a healthier lifestyle?" and made teens think towards their future health goal. Each teen had different goals depending on their problems revealed through comparison maps. Some of the participants had a healthy goal of doing everyday breakfast; another one had a goal of an improved body image by gaining muscles, and another one had a goal of evaluated their free time in a more productive way. Two weeks after the study, the researcher followed-up with teens regarding their future goals, whether they started to change their behaviors or attempted to start a change in their life towards healthier habits.

4.4.6 Results

Teens logged their casts on the mobile application for one-week. Some teens were not able to upload their videos due to lack of Internet connection through Wi-Fi or technical problems with their personal computer. The lack of SIM cards and Internet connection did not allow some videos to be uploaded in real-time to the web platform. The alternative solution was organizing another workshop to upload all videos and encourage them to interact on the website.

The workshop included one and a half hours discussion to get feedback from the one-week logging experience and the problems they face with, uploading all the content to the website, teach teaching how to comment, like and finally viewing videos, giving each teens two tasks on how to use the website. This alternative solution also gave time for nutritionist, counselor and teens to go over all of the videos uploaded on the website before the final workshop. One of the results from this workshop was that teens do not like to do tasks outside the workshop. Therefore one additional workshop and assignment have been added in the iterative cycles of the deployment phase.

During one week, teens had seen all videos on the website and were able to start exploring and interacting with each other. During the deployment time, the researcher posted "Summary of the day", a video series on the social network to summarize teen's work. These videos to motivate, inspire, and stimulate their interest on the project.

Teens took "Summary of the day" videos as guidance to solve some technical problems that they encountered with their phone. The research has an important role of creating a relationship with the teens, not forcing them to do things, but finding a way to make them do it voluntarily. This especially occurs when the researcher encourages them to do tasks; such as on the website or record a community cast. The researcher should not continuously and routinely send them emails or text messages to remind them of things, but in an easy and offhandedly manner, should ask occasionally if they had sufficient time to do something of particular importance. One of the teens commented "It is boring if we have Gmail reminder in each meal".

Teens' behave differently in their relations with a counselor they like and trust. The counselor's aim should be to capture their behavior in the most natural state and to decrease the pressure of 'unseen forces' such as authority or variable friendship. In this case, recording a video by them gives the feeling of being more intimate and keeping some distance from the force of authority.

During the deployment, a team of technical researchers, principal investigator and designer dealt with a variety of problems. This type of project requires multidisciplinary teamwork where the principal researcher can focus on conducting the research to achieve research goals. It is important for the study to find higher amount of participation, where in most cases the amount of participation or involvement in the project actively decreases. This is especially crucial to consider while working with young teenagers as in Locast Health Diary, that out of six teens, four of them actively participated in all sessions; two of them partially participated without making a full effort.

Another important issue was to select the right means of communication with participants of the study. In particular with younger age target groups, it is not easy to communicate, and most of the time the communication occurs through a mediating person such as the counselor, teacher or healthcare expert. During the workshop session, it was observed that teens rarely check their emails; instead they "text", or send SMS or write from a social network to communicate with their friends. Therefore, these new forms of media could be more effective in communicating with both for the counselor and the teens for the rest of the workshop.

The researcher also has multiple roles in the study. One of these roles is to talk with teens and motivate them to participate in the study. Another role is contribute to scientific knowledge, define methods used to analyze the data, and how it effects the outcome of knowledge creation. This is very important, not only for participants but also for intermediate persons who may be involved. In a researcher's role, it is important always to coordinate and examine other stakeholder's willingness to stick with the program and to control the process to assure that those stakeholders or other participants are doing what has been agreed is right. Another important result from Locast Health Diary study is the analysis of videos needs to be improved due to excessive amounts of data collected, and the content could be narrowed down to one singular activity such as focusing only on snacks rather than eating, exercise and free-time activities. However it is also important to capture all activities and their relations from multiple perspectives than going into detail of one singular activity. Especially, if you do not have a good idea of the participant's personality or the need for assessment of each teen, it is difficult to generalize which activity would work better to develop awareness on their everyday life. However, narrowing down the activity could be an alternative solution if it is integrated into a full program. In this full program, where each week or session, one activity could be monitored each week or session, and by focusing on its problems and setting future goals on that particular activity. Each activity could be also defined after the personal need assessment of each participant.

Another issue to consider is the type of social network that the study has proposed to participants. During the study, there was no obligation of using the Locast social network; therefore it was not used as much as expected. It was a new application for teens and was not usual for them to check their profile on that platform everyday as they do with other used social networks.

Locast Technology Problems
During deployment, there were technical problems regarding Wi-Fi access for synchronization, time and location characteristics of the phones used in the study.

The first problem is related to the characteristics of mobile phones (location and time set up configuration), which relates to other mobile technologies such as location detection, through Wi-Fi or GPS. If the mobile phone cannot detect the right location, the application saves a wrong location; or if there is no Wi-Fi and in addition to that if the participant is in a closed space where there is no signal for GPS, a location cannot be saved. The phones that were used for our deployment did not have a very sensitive GPS, so the phones had a hard time getting location fixes indoors. This device was augmented by a cell network using the technique called a GPS or Wi-Fi-based positioning. The possible solutions for receiving poor GPS was to stand outdoors near the place that one is eating, or near a window, and to wait until the icon in the status bar (that looks like a satellite receiver with blinking lines) stops. The time stamp of the videos on the mobile phone caused a problem of synchronization to upload videos on the website. Since phones were coming from the UK, all phones were manufactured with a time set-up on Greenwich time, thus the system had problems in converting into Eastern Time. Due to not having SIM cards on the mobile phone, there was no continuous Wi-Fi connection. Locast system remembered and used the last-found location, which the devices were able to accurately get but which kept setting wrong geo-tagging for the videos. Some steps to help resolve this problem were to teach the teens how to turn on their GPS feature on their mobile, ignore the location, re-locate with the right location on the website.

The second problem was related to the matching of the system components. Mobile and web interface communicate through API, which serves as a pipe between these two components. Since Mobile application needs to communicate with Locast web and as a consequence to Locast core in order to save data in the database, it has configuration problems (System language, codes etc.).

Although the quality of videos and sound properties were quite good, both teens and experts found that the quality of sound and videos could be even better on the website and allow one to analyze and understand the details in the recorded environment. The background noise was sometimes disturbing, especially when recordings done in a social environment. The sound and video quality could still be improved on the website. Mobile phones record 5 MP, 2592 × 1944 pixels video recordings. When you record video on the mobile side it configures to the website by converting the right file size. The amount of lightweight file size needs to be configured from mobile to web transmission. It is a part of the conversion process that converts videos from mobile to a web-friendly format.

There were video processing and synchronization problems from mobile to server and from server to website, due to a system communication error between Locast Web and Locast Mobile application. Although some of the casts could be seen on the mobile side, they could not be seen on the website. The web developer had to re-processed the videos to fix the problem.

Usability Feedback with Discussion Group Questions

During Locast Health Diary deployment, some technical issues were encountered on the mobile and web application. The most concerned issue was the uploading and the synchronizing of the casts on the mobile application. On the mobile application, it was possible to synchronize only all the casts; however, one of the teens was concerned about the issue that it would be useful and less time-consuming if each cast could have been synchronized standalone without any need to go through all the synchronized casts.

One of the teens states, "I liked the quality of sounds that the mobile phone was able to capture and I also liked the speed of it. Even though I did not use the apps or other functions on the phone, I find them very cool. Something I didn't like was the uploading part of recording. It takes a while and even if there was Wi-Fi, it might not even upload at all. I think I would want the uploading process easier than it is now. It would also be cool if we can see on the screen of our recording while we're recording. Like those iPhone or iPod cameras that switches around the image so we can see how we look like."

The problem here was that both techniques required having a working network connection. It was possible to get one through Wi-Fi, in which case the phone would be able to figure out its location when it had a Wi-Fi signal. Unfortunately, there were not many solutions to this issue beyond having it ignore old locations or giving the phones a network connection.

User Experience

Teens felt very comfortable using the Locast technology. They were able to do tasks on the website easily and very quickly in learning technical functions of the mobile and website application. This is an advantageous way of deploying the study with teens since they are eager to be familiar with the new technologies. However, it was relatively uncomfortable for some teens to record their diaries in a social environment, in particular when they were with friends. One of the teen's feedbacks was "I didn't feel bored while recording, but I did feel uncomfortable recording when other people are around me at first. It was awkward in school."

Teens did not use the social network (Locast web platform) actively. Longer timeframe was needed to get them used to using a new social network. One of the results was that teens needed more time for deployment. Ideal time frame could be a week and a half, although constant recording was an issue during the study. Sometimes teens missed to record their activities: "Sometimes, I forget to bring the phone to school, sometimes I forget to record, sometimes when you have to do something, you don't. I don't want to feel that I need to do it." They were also not so convinced about receiving reminders all the time: "I think it is annoying, if it is all the time, constantly." There could be alternative options to integrate the reminder in the core application where teens can set up their own agenda and personalize it. There was an additional feedback from a female teen that the retro camera could be an optional function to record one self.

4.5   Evaluation

The six teens uploaded 65 videos on the Locast website (http://locast.mit.edu/healthdiary); 26 casts were in the "Meal" category, 9 casts were in the "Snack" category, 4 casts were in the "Exercise" category, 19 casts were in the "Free-time" category, 5 casts were in the "Speak your mind" category, and only 2 casts were in the "Community" category.

The number of casts recorded by participants varied depending on the age group and the personality of the participant. 20 casts were the highest number recorded by one participant, where 4 casts were the lowest number. Two participants were 18 years old and they were 12th grade senior high school students; another four were 14–15 years old, 9th grade high school students. The oldest mostly had a tight schoolwork and exams. They recorded their casts more completely and created a more constructive narrative; three participants mostly did not speak much or in an incomplete way, and recorded fewer casts than others. There were freshman versus senior teens. Seniors were mostly busy with school schedules where freshman commented that Locast sounded easy and fun.

The method of analyzing videos had various methods where teens and researcher compared all three maps to draw results from the study. In the discussion sessions, teens compared their initial idea, which was logged in 'Typical Day' map on the 1st workshop, and the 'Locast Day' map, which was logged during the study. Researcher compared their 'Locast Day' map and their 'Future Goal' map to see whether they decided to change behaviors or not.

After analyzing all three state of cognitive and visual maps (Typical day, Locast day and Future goal day), video diaries, workshop discussion and post project questionnaire, researcher analyzed the Locast Health Diary study asking whether Locast tools and study methods created awareness on teens or whether they think to change (Fig. 4.11).

The comparison between "Typical day" and "Future Goal day" in visual and cognitive maps compiled by each participant showed that all participants developed awareness on healthy habits but two of them did not want to change their behaviors in practice, although all of them stated in the after-project questionnaire they would want to. Most of the participants developed awareness on eating and exercise and thought they could change in a positive behavior (such as eating more vegetables, having breakfast every day, not eating any snack two hours before sleeping, burning more calories after school activities, and drinking less coffee).

Most teens thought to change their behavior by adding a new activity (instead of surfing on the web, start playing a musical instrument, or instead of not doing any exercise, start playing volleyball) or changing of an actual habit (having breakfast earlier in the morning before class, exercising more after school) or decreasing the amount of an unhealthy habit (decreasing the amount of chips or candies eaten as snack in their Future Goal day) in their Future Goal map.

Developing awareness does not necessarily improve teens' behavior. For example, one participant developed awareness during Locast day that he had an

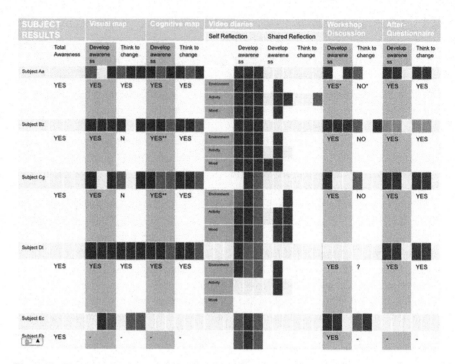

Fig. 4.11 The analysis of "Locast Heath Diary" study

unhealthy eating habit and he stated that he would like to continue this awareness in his "Future Goal day". On the other hand, two participants did not create awareness on free-time activities and actually wanted to change their behavior in a wrong direction, increasing the amount of non-social free-time activities. (One participant thought to add playing video games as free-time activity in his "Future Goal" map.) Most teens did develop awareness less on free-time, more on eating and exercise activities (Most of the male teens stated that sleeping was a free-time activity).

According to the post project questionnaire, they all developed awareness more on eating than exercise and free-time activities. Some of the teens developed awareness through self-reflecting on their own video, where others looking at other teens' diaries. In video diaries, they mostly created awareness of eating activities considering all aspects of environment, activity detail and mood.

All participants except one agreed that they felt comfortable recording their eating and physical activity. Most participants agreed that they felt more comfortable recording their eating activity than physical activity. Two participants somewhat agreed that they felt comfortable seeing their own video or other teens commenting on their video on the website. All participants agreed, when viewing other people's diary, that they became more aware of their eating and exercise habits. Except one participant, they also became more aware of their eating and exercise habits when viewing their own diary on the website. Three participants agreed that their perception changed through using mobile diaries and through social interaction on the platform. One participant strongly agreed that perception of exercise changed more through using mobile video recording then through social interaction. All participants agreed that their perception of eating changed through social interaction on the website and using the mobile recording of their activities. However, they agreed that their perception of exercise changed more through social interaction on the website than using mobile recording. All participants liked the workshop as a self-discovery experience. All participants agreed that they would change their eating habits and be more active after this workshop, where all agreed that the participatory workshop made them more aware of a healthier lifestyle. Analyzing the social interaction on the website, the results showed that viewing others' video diaries developed awareness and that commenting on each other's diaries provided interaction. As an example, all participants agreed that videos tagged with 'Eating dinner with family' were a social and healthy behavior than eating in a fast food restaurant or take-away food.

Teens comparing their "Typical day" and "Locast day" map developed awareness of their everyday behaviors (such as watching too much TV, skipping breakfast, eating fast food, snacking a lot, eating while not hungry, eating oversized portions, skipping meals), details in their co-related activities (such as snacking for fun with friends, studying, watching TV, sports game, surfing on web and snacking, emotional attachment to particular snacks, food type (i.e., rice, candies) and drinks (i.e., smoothies, soft drinks, energy drinks).

In the discussion group, one participant stated that viewing one of the participant's video in which he was always eating fast food discouraged her from eating fast food. During a discussion session, Locast diary tools provided discussion topics such as: if certain types of food (coffee, gums or multigrain bagel) are healthy or unhealthy; the reason why they skip breakfast; or why they choose a particular snack from a particular store, were among the topics discussed by the participants and experts.

Location-based maps were useful for researcher and experts to estimate teens' daily choices in their community. For example, one teen always ate his lunch in a fast food chain outside school, due to its availability and proximity to his school, and because he thinks that school lunch is not healthy and tasty.

Privacy is an important issue for social interaction on a website. According to the post project questionnaire and discussion group, some teens stated that at first they felt uncomfortable recording their activities, although they felt more comfortable once they had seen their own video on the website. Motivation is another factor to be considered, while most participants agreed that technology was innovative and interesting to them, although they did encounter error and problems with the technology.

Consequently, the results of the workshop showed that Locast Health Diary helped developing awareness. However, without expert participation it may not be sufficient to determine behavioral change.

Discussion groups and expert interviews demonstrated that Locast Health Diary could be an effective instrument for nutritionists, youth counselors and healthcare consultants in different ways. Medical experts see video diaries as a complementary tool useful for example to really get an idea of portion sizes and environmental factors. Educational experts see Locast Health Diary as a discussion tool able to monitor co-related activities and are interested in an extension of Locast Health Diary as a part of obesity prevention programs and its application to other wellbeing projects. This is a relevant result of the project since further refinement of action research according to Jacobs et al. (1992: 431) is that the results obtained from the research should be relevant to the practice.

Group Discussion

Group discussions were focused on analyzing and evaluating, as a result self-reflecting and learning from the video diaries. Many discussions developed awareness on their level of knowledge of healthy living. Teens thought that eating fruit and vegetables makes a healthier lifestyle. Although part of that statement is correct, on the other hand it is not enough to assure a healthy life. The success of a healthier lifestyle depends on balancing all activities and emotions in the long term. If it is only considered from one side, ignoring the other side, it does not mean that you have healthier habits and choices. Each activity and mood relates to each other, which defines our general wellbeing. Teens compared their "Typical day" map, their perception of their everyday activity schedule activities with "Locast day" map, six days of their reality of recorded activities. Experts and the researcher directly observed and helped teens directly while they evaluated their activities on a map.

> Teens did not record all their eating activities throughout the day, in particular the evening recordings. They did not record the food that they ate after 8PM, and some teens under-estimated their meal time until they check their Locast map. One teen mentioned that she paid more attention what she ate, "*Yes, it is like, you are putting it online, you have to at least speak more and eat healthy, at least try to do something.*" In some situations, teens were already aware that eating less was unhealthy but they accepted the situation and did not take any action to change it.

In teens' diaries, it is seen that their meals are in controlled portions. Locast helped teens to look back on what they had done throughout the week, and let them evaluate them and reflect on what they did.

Viewing another teen's diary changed the vision of some teens. They not only emphasized the bad situation but also they did not like to do that activity either.

> One of the teen stated criticizing his friend's video diary, "*When I was looking at Gavin's video, it is all about fast food. It is all about, that and I am like aghhh. It did the opposite effect I did not like it. I also commented on his video, 'fat is gonna stuck in your blood vessels.*"

Some of the teens are aware of what they are eating as in the conversation between teen counselor and Gavin; however, they do not have any other available option in his mind to make his lunch decision in a healthier way. In this dialog, we can also understand that the teen does not like the school food, and this situation brought him to eat in a fast food restaurant.

Expert asks Gavin, *"So Gavin how did it make you feel when you look back and see how much fast food you consume?"*

Gavin responds, *"At school we are all off in campus, we cannot go out. Fast food is better than eating the school food cause you don't know what they put in it these days, all it is chemical staff."*

In this dialog, we can see that he does very healthy activities but he does not show to other teens that he is doing these activities. Motivation and trust are two crucial issues in involving teens to record their everyday activities.

Expert adds, "One of the good side of Gavin he bikes everywhere he goes."
Gavin continues, "I bike at least 2 miles a day"
Researcher asks to Gavin, "I did not see any videos."
Gavin responds, "Because the time you guys gave me a phone, I had to do lots of staff, so I couldn't record."

Co-related activities are important behaviors that need to be analyzed. Watching a hockey game at a playing field is a behavior better than sitting in front of a TV, where you do not move. However, still it is also a sedentary activity where most of the time it is accompanied with snacks.

Teens tend to have hobbies that they do not do anymore, or some of the teens do not have time to do any exercise due to busy school schedule. Some of the male teens tended to joke about their answers on eating unhealthy food or pretending not to eat healthy food, justifying the real fact that healthy ones are more expensive than the processed ones.

Most teens' choose to eat what is convenient and available to them, as well as in some situations they also want what they do not have available at home.

Weak and Strong Points Map
Teens compared their "Typical day" and "Locast day" map and wrote down their weak and strong points.

The researcher gave examples of weak and strong points to help teens to evaluate and categorize their behaviors, such examples as "waking up at 6AM in the morning is a strong point, which means you are consistent to wake yourself up in the morning; or watching too much TV is a weak point which you perform sedentary activities too much of the time."

Teens put their thoughts next to each other and started thinking about what they would ideally like to change to create their future goals for a healthier lifestyle. They realized that they should be eating breakfast every day; they wished to work out more. While teens were creating their "Future Goal day", the counselor added comments to make them think about more realistic goals.

Expert Analysis of Video Diaries
Nutritionist evaluated teens' health diary videos in eating, free-time, and physical activities through "Video Analysis Map".

Analyses of Video Diaries | Eating Activities

Zhen's Diary: Zhen always expresses her feelings and she is the most expressive of all teens. She used 'speak your mind button' often to explain her mood during or after activities and missing activities throughout the day. She has a tight schedule at school therefore mostly working, doing homework and writing essay in her free time or during meals. She is mostly snacking while doing co-related activities. Although when she is snacking she is not hungry and she is mostly doing schoolwork and snacking at the same time. She mostly records her videos while she is alone at home, rarely at school breakfast in the classroom. As she also mentioned she felt not comfortable to record among friends. She never did any exercise, but one day she records her entire day what she had done, what she had eaten during the day through speak your mind button, where she meet with friends outside and went shopping. She always has family dinner with lots of home cooked food where they watch Korean TV.

Adam's Diary: Adam seems comfortable recording his diaries in social environment unlike other teens.

Adam is mostly in a social environment with friends at school and outside. He exercises a lot and spends lots of energy and as a spirit competitive and excited while he is playing Ping-Pong with Gavin. He says he feels tired all the time. He is all the time out and he is mostly consuming take-out or restaurant food. He is complaining about the school food "I am hungryyy. My friends are eating but I am not eating because the school food sucks." In one of his cast 'Dinner' he mentions that "I am not hungry but it looks good."

Nutritionist analyzes Adam's diary with the following words, *"Often states that he isn't eating because he isn't hungry—but all of his friends around him are eating. Is there a deeper reason as to why he isn't eating? Breakfast sandwich with orange juice one day for breakfast—looks like sausage-could replace with leaner meat like ham. English muffin is better choice than bagel for portion control. Tropicana orange juice was great choice-100 % juice! His diet seems to be mostly carbs—bread, noodles etc. Often times I couldn't see much of what he was eating—would need to have more one on one. Often surrounded by friends when eating–and usually away from the home—decreases chances of eating home cooked meal at home with family. When he does eat at home, seems like much healthier choices."*

Gavin has a speech disorder while he was young, therefore has already difficulties to speak in social environment. Therefore he records his casts at school or with friends are without any narrative; however while he is alone, he expresses himself very well. This tool might also open opportunities for people who might have speech problems. He is mostly choosing fast food options and substituting meals with unhealthy snacks. Like Zhen, he studies for S.A.T exam, therefore he snacks while he is studying. Gavin bikes every day to school; even he did not record this activity in his diary.

Tony's Diary: Tony has uploaded 4 casts, in which 3 of them were under the category of Snack and one was under the category of Meal. 3 out of 2 casts related to Snack casts were in a social environment with friends. He explains the portions. While he has co-related activities while snacking, he is surfing on web. He is snacking together with friends while they are talking and having fun. All his snacks were processed snacks such as bagel, cream cheese, and chips with dip and smoothie juice. While he is snacking he is in a good mood and he is not hungry. He states in one of his Snack video "Am I hungry? No, I am not. I am all right as always. I did not buy the snack, it belongs to my friend." He surrounds by his friends who snacks unhealthy food while socializing. While estimation; in one of his Snack cast 'Naked Snack', he estimates the ingredients looking at the nutrition facts; reads the information on the packaging; however it was difficult for most of the teens to analyze what does it mean and, seems complicated that they sometimes only show it in a short time.

Nutritionist analyzes Tony's diary with the following words, *"Bagel with honey and "cream cheese" frosting as a snack. Oversized portion for a snack. Could have had half and chosen a better topping like peanut butter or just the honey. Chips and salsa as snack—ok sometimes but shouldn't be every day. Better choice would be popcorn. Naked juice drink with protein—good! Better choice than soda. Good way to get fruits into diet if not consuming enough. But shouldn't replace whole fruits if given the option."*

Nutritionist analyzes Hieu's diary with the following words, *"Lunch: rice and meat was good. Major issue: 2 extremely large gatorades—consuming 8 servings rather than 1. About 400 calories of sugar and nothing else. Plain water would be a better choice."*

Chen's Diary: Chen has recorded four casts on "Eating", in which three of them was "Meal", and one of them was "Snack" substitute meal. She works part-time in a store, therefore she has more defined schedule. However in her "Typical day" map, it seems that she eats dinner very late like 10PM. Chen's diary has control portion meals and looks very healthy. However, it looks that the variety of food could be changed, including more vegetables and fruits.

Nutritionist analyzes Chen's diary with the following words, *"Generally speaking, the videos were hard to hear and to see exactly what was going on. Overall, when the teens ate at home, they seemed to make healthier choices in both the type of food and the portion sized because the meals were home cooked with fresher, healthier ingredients like meat and veggies and rice. When eating away from home, the portions were much larger and not nearly as healthy: oversized bagels, hot dogs, pizza, chips, fried foods, hamburgers and French fries. I didn't see much fruit being consumed, if any at all. Also didn't see much recording of the liquids being consumed. Water should be recorded just as soda, juice, and milk is. Need to incorporate more focus on fruits and veggies!"*

Analyses of Video Diaries | Exercise and Free Time Activities
Most of the participants had little idea how to assess this free time in a more constructive way; rather they stayed at home and socialized with friends through social networks or surfed on the Internet. Mostly teens, they did not have productive, creative or expressive free-time activities. Most of them surfed on social media channels, met with friends, and watched TV in their free time where they tended to snack in most of their sedentary activities.

As the nutritionist comments, *"Generally speaking, the more someone is sedentary, the more tired and lazy they will feel. For example, if someone sleeps too much, they will most likely wake up feeling groggier and lazier. People have the tendency to eat more when they are tired, bored, lonely, or just sitting around the house watching TV or on the computer. So if these participants go to school and sit all day in class and then go home and sit in front of the TV or computer, they will most likely feel lazy and tired and have little motivation to exercise. In addition, poor food choices like chips, candy, soda, and fast food make people feel tired because they give quick energy that wears off quickly. This is why fruits and veggies are better choices because they provide long lasting energy."*

"In regards to exercising, it is unfortunate that in Massachusetts we don't get much time to be outside and to exercise. If the students joined together and did exercise as a group or joined a team they would be more inclined to exercise because everyone around them is. And is makes it fun. The same thing goes with food choices. If a student is surrounded by friends who eat junk food or fast food all the time, most likely they will choose similar things in order to fit in with the crowd. If teen is more inclined to eat at home, where the family cooks their meals, then the choices will most likely be healthier. This isn't always the case but the chance is higher."

Male teens mostly exercised more than female teens; where during the week neither of the female teens did exercise. The most common physical activities were going to gym, biking, playing table tennis and basketball.

Analysis of Social Interaction Patterns

Commenting was the most important interaction pattern on the website. Teens commented on their video and on other teens' videos if they found the video content to be a healthy behavior or not. They liked, viewed and commented on each other's videos.

Zhen has the most constructive comments. Although some teens would like to only show their healthy habits, during discussion session they have admitted that they did not record some unhealthy habits.

Teens evaluated their own videos tagging with "No idea", "Unhealthy", "Healthy"; and voted on the other teens' diaries tagging with "The healthiest free time", "The healthiest exercise", and "The healthiest eating".

Some of the comments on the Locast website encouraged teens to interact with each other. Teens commented on videos and discussed the notion of having healthy/unhealthy habits. One of the comments to Gavin's lunch video was, *"Don't eat tooo much:). Fats are gona stack up in your blood vessels"*, discouraging him with chronic medical conditions from eating fast food.

Teens were defining what is a healthy sport in Adam's free-time video, *"Basketball is healthy, makes you work out and makes you tall."*, or evaluating what is unhealthy, *"I love 'Watching TV criminal minds' show but I have to admit that it's unhealthy cause we sit around wasting time and getting fat."*

All content published on the website was analyzed by the web administration, and were taken under control for any published content, which is not appropriate. Privacy and security issues are regarded as important factors to maintain consistency in the network.

Location-based Assessment
Location-based platform helps teens to map out all their activities in the community. This type of platform provides information about the relation between location of the activity, the teen and content of the video. The map gathers more information about the location of the activity, and the relation between locations of different or same type of activities. Experts can utilize this platform as a complementary tool to analyze teens' diary.

Teens' activities distribution all over the city shows the common places that teens go to eat exercise and spend their free time. Through location-based platform, it is possible to filter single types of activities and extract meanings from the analysis of this activity and its distribution in the city. This can be applied to analyze only one teen's map, or all teens' same type of activity map.

As you can see in (Fig. 4.12) the location of Zhen's activities are centered on home and school, on the other hand Gavin's activities locate in different places of the community. This information could tell us that he was a more active person than Zhen during the one-week time. The location of Gavin's lunch activity and the location of Gavin's school are very close to each other. Even though there are available other restaurants around the school, he preferred to eat at a fast food chain most of his lunchtimes. One of the reasons for this is the convenient price as well as the addictive taste of food. This type of sorting out activities based on locations could give us information about the availability of places around a certain spot.

Expert Evaluation through Interviews
Experts' evaluation is very important to realize if teens were aware of their problems and especially if Locast help them to develop this awareness and by which tools it developed awareness. In an interview session, the counselor, the nutritionist and the wellness consultant discussed the potential applications of Locast deployment and how could be integrated in the community programs.

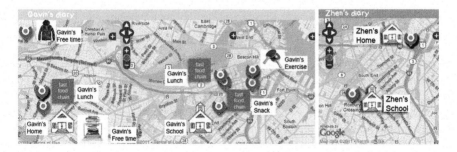

Fig. 4.12 Locations of two participants' diary in the community

The project results have been reviewed by experts and have opened new collaborations. The teens' counselor proposed that Locast application could be applied as a longer-term program in Asian Youth Essential Center. The healthcare consultant at Tufts Community Wellness thought that this could be applicable to other chronic disease conditions. The clinical director and nutritionist who works with obese teens at Center for Youth Wellness at Floating Hospital for Children would like to be part of a research proposal for grant Locast Application, where they can provide participants for the program. The Welcome project, Somerville, program director would like to apply Locast for other types of wellbeing projects involving video voices of physical activities of community members.

It has been understood that Locast Health Diary can be a discussion tool for educational experts, an additional food diary tool for nutritionists, an opportunity to implement different health protection, and community related context for community leaders. The Locast Health Diary develops awareness, however it cannot change behavior without an expert.

The nutrition educator and health coach of YES uses traditional tools to prevent obesity. Currently the nutritionist of YES group teaches nutrition to grades k-8 in Somerville. She states, *"In my 7th and 8th grade classes, I have them keep a food journal for 2 or 3 days a week. I encourage them to do a weekday and a weekend day to compare the differences when they are at home, school, etc."*

She thinks that the video diary is a great idea and is a fun way for the teens to do this project, and states "I think they should also write out their food diary, as it is a great resource for them to see what they eat and to reflect on ways they could improve or to notice positive choices they made". She thinks that Locast system could definitely be a tool to confront the obesity program but it would take much commitment, such as carrying around a phone and recording everything you do which can be stressful and tiresome. She does not think that Locast system alone could prevent obesity, but definitely creates awareness and self-reflection for the person doing the diaries and for the people who are watching them. She also thinks that it should involve a written food diary portion, which makes it easier to keep track of and to reflect on again. She added that the only thing is that there is an increased tendency to lie when people write down their foods because they often feel guilty if they make poor decisions. Therefore she thinks that Locast Health Diary tools should be used in addition to the written diaries where the video will help to really get an idea of the portion sizes and the environmental factors.

David Vo, youth counselor from Boston Asian Youth Essential Service thinks the program itself has potential to help teens be more aware of themselves on what they are doing. He states, "One of the key things to get this into a better position is more of a 'one on one' at first talk, this is how you do it versus giving them and find out, we just want the truth". He adds that the definition of the project goal at the beginning that what is really help them, see and focus what they are doing to help them in this project.

According to youth counselor, Locast system definitely does have a potential tool to prevent obesity. But he thinks that it also a lot depends on the person you are dealing with. He states, "Because of the group that we are working with, you need to deal with personality. Some of them are outgoing and some of them are open, it is almost like a therapy session, taken for kids to open up. The thing about Locast is definitely a tool to align the work of young people. But to use it as the only tool is hard since we work with the personality." As an example, one of the girls' was very self-conscious on what she is eating, however she was not very comfortable to share with others that she was eating fish every day, which in Asian culture it can be normal to eat fish every day as opposed to other cultures. On the other hand, one of the teens was an active contribution, and wanted to show all his activities from 6am to 6pm. Another different personality example was one of the teens had a speech disability problem which video diaries was a way to give him self-confidence of his speech problem. It is important to understand and assess participants' personality, ethnicity, and culture to customize Locast Health Diary tools tailored to their problem.

The youth counselor says that they offer services that come in different shapes. He states, "*For an individual, they first define and assess their needs, and figure out what are the needs and resources do we have to fulfill those needs. So as far as Locast program is concerned, definitely good way to keep up with video diaries aspect/technology, but the doubt about that is that kids now they are too dependent on technology and video diary, so there is not very much one to one contact. Us, here at YES, youth counselors and youth workers we are always trying to help kids learn social skills; to walk into shake someone's hand, to keep eye contact. I think nowadays youth are too dependent on technology, sitting behind the screen, instant text messaging. I think that where there might be some both could help each other out but also hurt each other. Where the hardest part for us counselors is to sit down with an individual and make him talk. Some kids are open to talk in front of a camera but some are not.*"

To integrate Locast Health Diary tools and methods into a program, he thinks one of the key issues is spending more time, how to use it, and what is the actual goal, what they are trying to accomplish. He states, "A lot of times you give kids camera, they just do whatever they do. Locast Health Diary tools would definitely goes into our toolbox. It would be beneficial for the teen's development that you are working with. This is a big project. 4 days is a kind of short time for teens to get used to recording audio-visual environment. It would be good if they could play with the phones for 3–4 days until they feel comfortable recording, and using video recorder. Even in my photography classes, I give them camera for 2–3 days a week, and they take random photos, after 2 weeks time they start learning to change angles, compositions even later to think about the light etc.… it takes time to learn and practice."

Michael Leidig, nutritionist at Tufts Floating Hospital, is very eager to use Locast Health Diary tools, and he thinks it is extremely important for some of the kids he works with to test this program. He states, *"They have unhealthy habits that make already make them into trouble. They probably need it more than you do. YES teens help them with this deployment. You have done a very important service. This pilot program and your working is very helpful for us, Locast we use it for different things for learning that's why we take pictures cause it is important what you do."*

References

Allison, D. B., Fontaine, K. R., Manson, J. E., Stevens, J., & VanItallie, T. B. (1999). Annual deaths attributable to obesity in the United States. *JAMA, 282*(16), 1530–1538.

Arslan, P., Brunnberg, L., Casalegno, F., & Schladow, Z. (2011). Locast H2flow: creative learning tool for participatory urbanism. *DESIRE '11 Procedings of the Second Conference on Creativity and Innovation in Design* (pp. 267–270).

Bibbins-Domingo, K., Coxson, P., Pletcher, M. J., et al. (2007). Adolescent overweight and future adult coronary heart disease. *New England Journal of Medicine, 357*(23), 2371–2379.

Boutelle, K. N., Cafri, G., & Crow, S. J. (2011). Parent-Only Treatment for Childhood Obesity: A Randomized Controlled Trial. *Obesity, 19*, 574–580. doi: 10.1038/oby.2010.238.

Clark, C. M. (2011). *Relations between social support and physical health middle-aged adults.* Corey M Clark in Health San Francisco.

Crandall, C. S., & Moriarty, D. (1995). Physical illness stigma and social rejection. *British Journal of Social Psychology, 34*, 67–83.

Crandall, C. S., D'Anello, S., Sakalli, N., Lazarus, E., Wieczorkowska, G., & Feather, N. T. (2001). An attribution-value model of prejudice: Anti-fat attitudes in six nations. *Personality and Social Psychology Bulletin, 27*, 30–37.

Csikszentmihalyi, M., & Larson, R. (1987). Validity and reliability of the experience-sampling method. *The Journal of Nervous and Mental Disease, 175*(9).

Delva, J., Johnston, L. D., & O'Malley, P. M. (2007). The epidemiology of overweight and related lifestyle behaviors: Racial/ethnic and socioeconomic status differences among American youth. *American Journal of Preventive Medicine, 33*(4S), S178–S186.

Dietz, W. H. (1998). Health consequences of obesity in youth: Childhood predictors of adult disease. *Pediatrics, 101*(3 Pt 2), 518–525.

Ebbeling, C. B., Pawlak D. B., Ludwig D. S., (2002). Childhood obesity: public-health crisis, common sense cure. *Lancet, 360*(9331): 473–482.

Fikkan, J., & Rothblum, E. (2005). Weight bias in employment. In K. D. Brownell, L. Rudd, R. M. Puhl, & M. B. Schwartz (Eds.), *Weight bias: Nature, consequences and remedies* (pp. 15–28). New York, NY: Guilford Press.

Flaherman, V., & Rutherford, G. W. (2006). A meta-analysis of the effect of high weight on asthma. *Archives of Disease in Childhood, 91*(4), 334–339.

Frayling, C. (1993). *Research in art and design.* Royal College of Art Research Papers 1, vol. 1, pp. 1–5.

Fontaine, K. R., Redden, D. T., Wang, C., Westfall, A. O., & Allison, D. B. (2003). Years of life lost due to obesity. *JAMA, 2003*(289), 187–193.

Garner, R. E., Feeny, D. H., Thompson, A., Bernier, J., McFarland, B. H., Huguet, N., et al. (2011). Bodyweight, gender, and quality of life: a population-based longitudinal study. *Quality of Life Research*

Goodare, H., & Lockwood, S. (1999). Involving patients in clinical research: Improves the quality of research. *BMJ, 319*(7212), 724–725.

Gordon, A., Crepinsek M. K., Nogales, R., & Condon, E. (2007). School Nutrition Dietary Assessment Study-III. Vol. 1. Princeton, N. J.: Mathematica Policy Research. School Food Service, School Food Environment, and Meals Offered and Served.

Gortmaker, S. L., Must, A., Perrin, J. M., Sobol, A., & Dietz, W. H. (1993). Social and economic consequences of overweight in adolescence and young adulthood. *New England Journal of Medicine, 1993*(329), 1008–1012.

Green, L. W., George, M. A., Daniel, M., Frankish, C. J., Herbert, C. P., Bowie, W. R., et al. (1995). Study of Participatory Research in Health Promotion (pp. 43–50). Royal Society of Canada, Ottawa, Ontario.

Grinter, R. E., & Eldridge, M. (2001). 'y do tngrs luv 2 txt msg?'. In: W. Prinz, M. Jarke, Y. Rogers, K. Schmidt and V. Wulf (Eds.), *Proceedings of the Seventh European Conference on Computer Supported Cooperative Work ECSCW '01, Bonn, Germany* (pp. 219–238). Dordrecht, Netherlands: Kluwer Academic Publishers.

Himes, S. M., & Thompson, K. J. (2007). Fat stigmatization in television shows and movies: A content analysis. *Obesity, 15*(3), 712–718.

Jacobs, C. D., Haasbroek, J. B., & Theron, S. W. (1992). Effektiewe Navorsing. Navorsingshandleiding vir tersiêre opleidingsinrigtings. Geesteswetenskaplike komponent. Pretoria: Universiteit van Pretoria. Effective Research, Research Guide for tertiary training. Humanities component, University of Pretoria.

Karremansa, J. C., Stroebeb, W., & Clausb, J. (2006). Beyond vicary's fantasies: The impact of subliminal priming and brand choice. *Journal of Experimental Social Psychology, 42*(6), 792–798.

Kimm, S. Y., & Obarzanek, E. (2002). Childhood obesity: A new pandemic of the new millennium. *Pediatrics, 110*(5), 1003–1007.

Kopelman, P. G. (2005). *Clinical obesity in adults and children: In adults and children.* USA: Blackwell Publishing.

Koplan, J. P., Liverman, C. T., & Kraak, V. I. (Eds.). (2005). *Preventing childhood obesity: Health in the balance.* Washington, D.C.: National Academies Press.

Lobstein, T., Baur, L., & Uauy, R. (2004). Obesity in children and young people: A crisis in public health. *Obesity Rev, 5*(1), 4–85.

Maranto, C. L., & Stenoien, A. F. (2000). Weight discrimination: A multidisciplinary analysis. *Employee Responsibilities Rights Journal, 2000*(12), 9–24.

McGinnis, J. M., Gootman, J. A., & Kraak, V. I. (2006). *Food marketing to children and youth: Threat or opportunity?* Washington, D.C.: Committee on Food Marketing and the Diets of Children and Youth: Institute of Medicine. The National Academies Press

Miller, J., Rosenbloom, A., & Silverstein, J. (2004). Childhood obesity. *Journal of Clinical Endocrinology and Metabolism, 89*(9), 4211–4218.

Moore, L. V., & Diez Roux, A. V. (2006). Association of neighborhood characteristics with the location and type of food stores. *American Journal of Public Health, 96*(2), 325–331.

Morland, K., Wing, S., & Diez, R. (2002). The contextual effect of the local food environment on residents' diets: The atherosclerosis risk in communities study. *American Journal of Public Health, 92*(11), 1761–1767.

Neumark-Sztainer, D., & Haines, J. (2004). Psychosocial and behavioral consequences of obesity. In J. K. Thompson (Ed.), *Handbook of eating disorders and obesity* (pp. 349–71). Hoboken, NJ: Wiley.

Ogden, C. L., Flegal, K. M., Carroll, M. D., et al. (2002). Prevalence and trends in overweight among US children and adolescents, 1999–2000. *Journal of the American Medical Association, 288*(14), 1728–1732.

Ogden, C. L., Carroll, M. D., Curtin, L. R., McDowell, M. A., Tabak, C. J., & Flegal, K. M. (2006a). Prevalence of overweight and obesity in the United States, 1999–2004. *JAMA, 295* (13), 1549–1555. doi:10.1001/jama.295.13.1549

Ogden, C. L., Carroll, M. D., & Flegal, K. M. (2006b). High body mass index for age among US children and adolescents, 2003–2006. *JAMA, 299*, 2401–2405. 2008.

Ogden C. L., Carroll M. D., McDowell M. A., & Flegal, K. M. (2007). *Obesity among adults in the United States—no change since 2003–2004. NCHS data brief no 1.* Hyattsville, MD: National Center for Health Statistics.

Ogden, C. L., et al. (2012). Prevalence of high body mass index in US children and adolescents, 2007–2008. *JAMA, 303*(3), 242–249.

Powell, L. M., Slater, S., & Chaloupka, F. J. (2004). The relationship between community physical activity settings and race, ethnicity and socioeconomic status. *Evidence-Based Preventive Medicine, 1*(2), 135–144.

Reason, P., & Bradbury, H. (Eds.). (2001). *Handbook of action research: Participative inquiry and practice.* London: Sage Publications.

Rich, M., Lamola, S., Amory, C., & Schneider, L. (2000). *Asthma in life context: Video intervention/prevention assessment, pediatrics* (vol. 105 (3Pt 1), pp. 469–77). Boston, MA: Division of Adolescent/Young Adult Medicine, Children's Hospital.

Rose, G., & Manderson, L. (2000). More than a breath of difference: Competing paradigms of asthma. *Anthropology and Medicine, 7*(3), 335–350.

Schön, D. A. (1983). *The reflective practitioner: How professionals think in action.* London, UK: Temple Smith.

Seagle, H. M., Strain, G. W., Makris, A., & Reeves, R. S. (2009). American Dietetic Association. Position of the American Dietetic Association: Weight management. *Journal of the American Dietetic Association, 2009*(109), 330–346.

Sechrist, G. B., & Stangor, C. (2005). Social consensus and the origins of stigma. In K. D. Brownell, L. Rudd, R. M. Puhl, M. B. Schwartz (Eds.), *Weight bias: Nature, consequences and remedies* (pp. 109–120). New York, NY: Guilford Press.

Sharp, E. H., Caldwell, L. L., Graham, J. W., & Ridenour, T. A. (2006). Individual motivation and parental influence on adolescents' experiences of interest in free time: A longitudinal examination. *Journal of Youth and Adolescence, 35*, 359–372.

Singh, G. K., Siahpush, M., & Kogan, M. D. (2009). Neighborhood socioeconomic conditions, built environments, and childhood obesity. *29*(3), 503–12. doi: 10.1377/hlthaff.2009.0730.

Speiser, P. W., Rudolf, M. C., Anhalt, H., et al. (2005). Childhood Obesity. *Journal of Clinical Endocrinology and Metabolism, 90*(3), 1871–1887.

Thompson, J. K., Herbozo, S. M., Himes, S. M., & Yamamiya, Y. (2005). Weight-related teasing in adults. In K. D. Brownell, L. Rudd, R. M. Puhl, M. B. Schwartz (Eds.), *Weight bias: Nature, consequences and remedies* (pp. 137–149). New York, NY: Guilford Press.

Trasande, L., Liu, Y., Fryer, G., & Weitzman, M. (2009). Effects of childhood obesity on hospital care and costs, 1999–2005. Health Affairs. W751–W760.

Troiano, R. P., Berrigan, D., Dodd, K. W., et al. (2008). Physical activity in the United States measured by accelerometer. *Medicine and Science in Sports and Exercise, 40*(1), 181–188.

Tuomas, L., & Oinas-Kukkonen, H. (2011). Persuasive features in web-based alcohol and smoking interventions: A systematic review of the literature. *Journal of Medical Internet Research, 13*(3), e46.

Uchino, B. N. (2004). *Social support and physical health: Understanding the health consequences of relationships.* New Haven, Connecticut: Yale University Press.

Wang, Y. C., McPherson, K., Marsh, T., Gortmaker, S. L., & Brown, M. (2011). Health and economic burden of the projected obesity trends in the USA and the UK. *The Lancet, 378* (9793), 815–825.

Wanless, D. (2002). *Securing our future health: Taking a long term view. Final report.* London: HM Treasury.

Young, L. R., & Nestle, M. (1998). Variations in perceptions of a "medium" food portion: Implications for dietary guidance. *Journal of the American Dietetic Association, 1998*(98), 639–641.

Zoumas-Morse, C., Rock, C. L., Sobo, E. J., et al. (2001). Children's patterns of macronutrient intake and associations with restaurant and home eating. *Journal of the American Dietetic Association, 101*(8), 923–925.

Chapter 5
Discussion

5.1 Towards Participatory Healthcare Services

The rapidly increasing prevalence of chronic diseases in particular obesity among
children and adolescents reflects a global 'epidemic' worldwide. These serious
medical conditions associates with hypertension, type 2 diabetes, metabolic syn-
drome, fatty liver disease, sleep disturbances along with greater risk of social and
psychological problems. We need a vision that integrates a lifestyle of healthy
habits with an environment that promotes healthy living by encouraging exercise
and making healthy food affordable. Prevention is highly important and there is an
urgent need for further practices in participatory healthcare services and community
health interventions using our advanced technologies. New technological frameworks engaging all key stakeholders to ensure discovery and delivery of these
technologies, and with the establishment of strategic long-term healthcare planning
will be crucial.

Participatory healthcare services will solve this problem in two levels: First one
is the involvement of individuals into their own healthcare management. Second
one is the participation of stakeholders of the healthcare system that enables a
healthier life services and environment for the individuals.

5.1.1 Involvement of Individuals

People are trapped in the current health systems; we need to provide best practices
to give them power and ownership in their own health decisions that affect a larger
community. As Wanles (2002) argues, the future of healthcare in an era of chronic
disease should turn on the full engagement of people in their own healthcare;
promotion of good health and prevention of illness. In order to create a healthier
lifestyle, patients need to manage long-term relations with their everyday activities

© The Author(s) 2016
P. Arslan, *Mobile Technologies as a Health Care Tool*,
PoliMI SpringerBriefs, DOI 10.1007/978-3-319-05918-1_5

towards a healthier life. This happens when the patient is informed how to cope with illness in the long run and is taught how to learn to live healthily.

Service Design provides participatory tools to involve individuals into their health management. Participatory tools could help individuals to participate, involve and capture their socio-psychological data in three ways: (1) by empowering people and giving them confidence and involvement in their cure process, (2) by involving the patient they feel more a part of the process and feel more ownership, (3) by giving them more control, they could be able to develop awareness on the right decisions for their health.

We are confident that, in the near future, diseases will be controlled and prevented by smart medical applications using software and hardware product services. People will have constant access to doctors, families and healthcare service providers anytime and anywhere. Biometric data will be monitored in real time by personalized secure devices that will provide easy access to anyone's personal health information. People and professional networks will easily share this data via cloud technologies. Smart applications will enable people to compare and discuss potential illnesses in an integrated healthcare ecosystem. All these advances are expected to lead to a greater awareness, participation and control of individual's health.

Due to remarkable advances in technology, the system we envision has the potential to recognize individuals and the individual does not need to ask anything about it. The physical meaning of objects that all become virtual systems changes the dynamics of social interaction. We use computers as tools to integrate the products we need to further our healthcare goals. Think of a city as a network of computers with a vast number of points of interaction. Our age of digitalization can be turned into successful social interactions to create communities, establish connections among them, and both motivate and support connections among individual citizens.

Locast Health Diary study aims to create a framework that will develop practices in different layers of the healthcare system, primarily through the use of mobile technologies to create social interactions. The objective is to create relations between health researchers and design and technological practitioners to strengthen the infrastructure of healthcare and to add new dimensions to the science of health practice.

It is not too far a stretch to imagine the use of mobile technology and civic media creation as a tool to engage individuals and communities in learning and modification of social behavior in healthcare. It will be necessary to understand the correlation of those behaviors with success over time in users' lives. Active and passive tools can be aggregated to this approach for a complete healthier life. A well-structured design framework can offer methods and tools to reflect on users' perception by connecting products, services and actors in the system. This will be an opportunity to understand the potential of technology and its relations between users for creating new scenarios in other healthcare contexts.

5.1.2 Involvement of Stakeholders

Knowing how to stay healthy is not enough to motivate individuals to adopt healthy lifestyles, but relevant progress can be achieved through the use of incentives delivered through a combination of processes and mobile technologies. Recognizing the effectiveness of this approach, further multi-dimensional cross-disciplinary projects will help individuals and communities motivate themselves on behavioral changes towards healthier lifestyles and prevent them from chronic diseases.

In future health prevention programs we should include all powers together. Advances in technology should serve as a tool to obtain various types of information on physiology, environment, and social behavior. The individuals' health depends on many influences in a complex ecosystem. One is dependent on another. The more data we have in each circle, the better predictions we could make for one's healthier decisions. There should be collaboration between advanced technology to collect data, participatory design to empower individuals and communities to do best practices, and finally if we combine these we would have better healthcare systems with proactive healthier life choices. Each individual's choice affects the other and spreads in the network.

As one of the main goals of this study is to shift the focus of our conversation from questions of technological access to those of opportunities to participate and develop the cultural competencies and social skills needed for full involvement. Schools as institutions have been slow to react to the emergence of this new participatory culture; the greatest opportunity for change is currently found in afterschool programs and informal learning communities. Schools and afterschool programs must devote more attention to fostering what we call the new media literacies: A set of cultural competencies and social skills that young people need in the new media landscape.

Service Design provides participatory tools (social networks, participatory workshops, focus groups, and open-ended discussions) to involve different stakeholders into the problem solving (Fig. 5.1). In a user-centered design approach, the designer tries to understand the system from a user's point of view, assuming therefore a user status. However, in a participatory approach, each actor in the system is a participant in the process, and a service designer needs to understand each actor's problems, and desires, and needs to create links to each other. As service designers, we now have a new competence in being able to think not only about users but also medical experts, policy makers, participants all of whom have different objectives and relations to each other. If the challenge is to solve a complex problem, it is important to give priorities to the point of interaction from which you would like to approach the problem.

Social networks are an important part of a patient's life in relating topics of "treatment" and "environment", two factors that act as parameters of health and wellbeing of one person. Rich et al. (2002) contend that the patient is a network of physical and psychological functions and interacts with physical, biological, social,

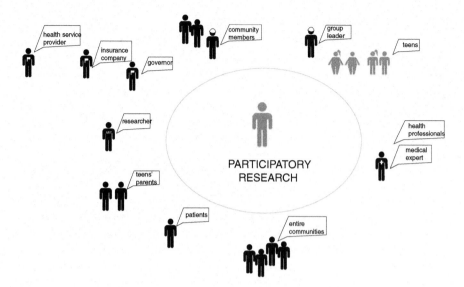

Fig. 5.1 Various stakeholders in participatory research

and symbolic factors. Designer can, and should, do more research on exploring how to participate, encourage and educate individuals to monitor and reveal the socio-psychological environment in co-designing research projects.

To sum up, the specific contribution of the service designer in healthcare is (i) creation of relations between actors through their creativity and intuitive competences, (ii) capacity to translate and transform ideas, desires, and the need of actors in visual representation forms. Visual tools are useful to communicate among actors, and (iii) developing new tools to help users monitor behavior. In a nutshell the problem to be addressed; for experts to help them structure, synthesize and communicate user behavior.

Advance the use of technology to better capture and analyze individual's socio-psychological environment and integrate with other quantified datasets for better decisions.

There are so many influencing factors and clues about patients' lifestyle outside the hospital, which are not regarded by medical experts and that can also contribute to manage patients' lives towards healthy behaviors. Social environments in which patients live should be a critical factor within the cure. Since the individual is the only person who can provide data about their 24 hour environment and health life, we should use technology in an advanced manner so that we can get accurate data from them.

Collecting and analyzing more socio-psychological information about a patient gives additional information for more complete chronic disease prevention. This type of patients' data is easier to understand and reflect by themselves respecting biophysical information where patients might need a certain expertise or help by an expert to decode information in their "everyday language." As Goodare and

Lockwood (1999) state "Only the patient knows about his or her experience of illness, social circumstances, habits and behavior, attitudes to risk, values and preferences." However, at the same time this is challenging since it is difficult to monitor someone's everyday life and it is totally dependent on their will power and their awareness. The issue which is even more challenging is how to change their behavior. We need to have successful long-term engagements with individuals: Education is one way, using technology tools to create or change behavior is another way, policy changes are yet another way. We need to think in a systematic way to solve this complex ecosystem and provide solutions at an individual level. It is crucial to improve video diaries in three aspects: (i) as a motivating and engaging tool for individual's life, (ii) as a technology tool to optimize the way we collect and analyze data, (iii) as an integrated method that connects with other methods to capture both physiological and environmental data to understand a complete picture of one's health.

5.2   Next Steps in Locast Health Diary Project

Locast Health Diary project is based on Frayling's "research through design" model (1993) and explores the practices and processes of design through participation. Design is a reflective practice where designers reflect on the actions taken in order to improve design methodology. The development of design practices is considered not as the objective of the research, but as an integral part of the project. This practice-based approach is a systematic inquiry with systematic reflections that occur in practice settings. The goal is to move the knowledge derived from creation to research. Therefore, further studies require that we assess Locast platform acceptability on a larger scale, usability and user satisfaction on a longer term implementation. This analysis will provide more accurate information in assessing an individual's lifestyle and behavior change.

As for the next steps to Locast Health Diary, the study was integrated as the core project into a European Union FP7-ICT funded work program with the objective of addressing personalized health, active ageing, and independent living (Carrino et al. 2014). The project relies on ICT technologies to implement a framework for the promotion of a health service based on three main features: a) individual and environmental monitoring, including wearable sensors, mobile phone and multimedia diaries for the acquisition of physical, physiological and behavioral attributes of participants; b) feedback to the user, presenting personalized healthy options for alternative lifestyles; c) social connectivity, encouraging involvement in social network experience sharing and social engagement. The study will be tested with over 300 adolescents in three European Union member states: Spain, Italy, UK. The development of the project will mobilize a wide stakeholders' ecosystem contributed by National Health authorities and research institutions, industries and Academia from the ICT and healthcare sectors, as well as food companies and small medium enterprises.

We need more experience with multi-dimensional and cross-disciplinary technological interventions that generate self-awareness (acknowledgement of risks associated to unhealthy behaviors), enhancing and sustaining motivation to take care of their health with a short/medium/long term perspective, and finally changing behavior towards a healthy lifestyle and becoming a co-producer of their health and taking an active role by sustaining it.

Cloud computing, and convergence towards the mobile platform as enablers are crucial points to create an integrated healthcare vision, including the three levels of the individual state at physical-physiological, nutritional, and psychological levels. The future direction on healthcare services using advanced technology is to provide guidance towards developing better capture and analysis of all states and to derive meaningful outcomes to deliver better decision mechanisms both for individuals and all healthcare system stakeholders. More research is necessary into the combination of technologies (wearable technologies, accelerometers, ECG, multimedia diaries, social networks, location-based platforms) with different stakeholder involvements in various healthcare settings, which provide intelligent systems for recognition of behavioral trends and early detection of health risks on the basis of this combined data, including data acquired by sensors and individual self-assessment.

The next generation's healthcare services should provide social platforms to stimulate individual's willingness to engage actively in their health management. This social platform also could enable contributions from all stakeholders of the health and wellness ecosystem, as the solution starting from preventive health can expand into a systematic approach to cover all levels of influence of individuals' health.

There are four main areas that advanced technology and participatory tools in service design could focus on research and implementation:

- Individual and Environmental Monitoring—This dimension consists of the environmental, behavioral and physiological analysis of young users, through a high level-monitoring platform including wearable sensors and mobile phones as well as multimedia diaries for the acquisition of physical, behavioral and emotional attitude of individual.
- Feedback System—Providing immediate real-time feedback in terms of "health status" changes, requiring actions that will undertake promotion of active involvement in changing their own behavior.
- Social connectivity and engagement—The third dimension extends to include a social network where the user can share experiences with a community of peers.
- Motivation and behavioral change—This is quite challenging since motivation depends on many factors as well as emotions, psychological environment and personality of teens. The system needs to provide constant different layouts of motivational activities. It is important to involve experts in the behaviour change process, who together with technology, can give guidance and monitoring over a longer period.

It is crucial to create uses of advanced technologies to better understand socio-psychological environments, and to quantify the data with other sensorial biological data with the aim of bringing more contexts into the healthcare decision mechanisms.

5.3   Future Directions

Wellbeing requires long-term management of our own health by engaging actively every day. This would bring less cost for healthcare and prevention of chronic disease, which aims to solve prior problems. Technology can help to manage and promote everyday health and social health connections enabling systems. Design provides new connections and offers new forms of participation to be applied in everyday context through new tools and methods. Scenarios around different chronic disease prevention through mobile technologies can be explored to promote healthier lifestyles and social connections through supportive environments and platforms that proactively help people remain healthy, autonomous, and engaged in life. The initial framework of research could be expanded to other stakeholders in the healthcare system; such as health insurance companies, other community initiatives, public health institutions and local organizations.

One of the future steps is to continue to tune into the contribution of service design research on participative methods toward healthcare prevention, and explore best practices in other healthcare fields such as diabetes, migraine, and asthma. It is important to keep a systematic view and a multidisciplinary approach to solve complex problems of healthcare. Although the healthcare research area intervenes multi-disciplines in its problem and solution; however, it is still lacking collaboration among different research communities in other scientific research fields. If these communities can be more open, then there will be an exchange of information and bilateral collaboration. This leads other communities to integrate design research to enrich this research area for a healthier future.

Research goes far beyond description of questions. It requires analysis, prediction and generalization, explanation, relationship and compensation through how and why questions. Design could provide and improve tools and frameworks bridging the gap between technology and health by creating social interaction services.

Further questions could be asked to intervene in different research areas to improve methods and provide scientific knowledge in theory: How design methods enable social and technological innovation for a sustainable healthcare and wellbeing context? How narrative approach and video diaries could prevent chronic diseases? How might we use mobile technology to improve social interactions for a healthier lifestyle? How can design practice be supported by research in order to generate technology-based relational innovation? What active and passive tools could you use in the context to create future scenarios around your perception of healthcare? How to engage and encourage users in their own long-term healthcare management?

Through Locast Health Diary project, it has been understood that it can be a discussion tool for educational experts, an additional food diary tool for nutritionists and an opportunity to implement different health prevention plans, as well as a community related context for community leaders. The tools and methods used in this framework have been useful not only for the participants, but also for other stakeholders' objectives for a systematic participatory health system. This system has been proposed by a clinical director at Child Health Center, Tufts University to implement for his obese patients, where teens really need a social motivation and a self-reflection on their behaviors. The same system was asked to be implemented in a community setting by a community leader in a different health promotion concept. This model could be taken as an insight for a framework and be implemented in a different context of healthcare prevention from a policy level to self-reflection level, and as a consequence could lead to a participatory, connected and accessible health for everyone.

Service Design Thinking **through mobile technology uses its intuition to see the hidden power of interactions and creativity to construct a healthier life for and by the people.**

References

Carrino, S., Caon, M., Khaled, O., Andreoni, G., & Mugellini, E. (2014). *PEGASO: Towards a life companion.* 16th International Conference on Human-Computer Interaction (HCII2014).

Carrino, S., et al. (2014). PEGASO: A personalized and motivational ICT system to empower adolescents towards healthy lifestyles. Innovation in Medicine and Healthcare 2014.

Frayling, C. (1993). Research in art and design. Royal College of Art Research Papers 1, vol. 1, pp. 1–5

Goodare, H., & Lockwood, S. (1999) Involving patients in clinical research: Improves the quality of research. *BMJ, 18*(319)(7212), 724–725.

Rich, M., Patashnick, J., & Chalfen, R. (2002). Visual illness narratives of asthma: explanatory models and health-related behavior. *American Journal Health Behavior, 26*(6), 442–453.

Wanless, D. (2002). *Securing our future health: Taking a long term view.* London: HM Treasury.

Index

© The Author(s) 2016
P. Arslan, *Mobile Technologies as a Health Care Tool*,
PoliMI SpringerBriefs, DOI 10.1007/978-3-319-05918-1

Printed by Printforce, the Netherlands